Lecture Notes in Business Information Processing 211

W0230425

More information about this series at http://www.springer.com/series/7911

David Aveiro · Robert Pergl
Michal Valenta (Eds.)

Advances in Enterprise Engineering IX

5th Enterprise Engineering Working
Conference, EEWC 2015
Prague, Czech Republic, June 15–19, 2015
Proceedings

 Springer

Editors
David Aveiro
University of Madeira
Funchal
Portugal

Michal Valenta
Czech Technical University in Prague
Prague
Czech Republic

Robert Pergl
Czech Technical University in Prague
Prague
Czech Republic

ISSN 1865-1348   ISSN 1865-1356 (electronic)
Lecture Notes in Business Information Processing
ISBN 978-3-319-19296-3   ISBN 978-3-319-19297-0 (eBook)
DOI 10.1007/978-3-319-19297-0

Library of Congress Control Number: 2015939232

Springer Cham Heidelberg New York Dordrecht London

Printed on acid-free paper

Springer International Publishing AG Switzerland is part of Springer Science+Business Media
(www.springer.com)

Preface

Enterprise engineering is an emerging discipline that studies enterprises from an engineering perspective. It means that enterprises are studied as being purposely designed and implemented systems. Enterprise engineering is rooted in both the organizational sciences and the information system sciences. The rigorous integration of these traditionally disjoint scientific areas has become possible after the recognition that communication is a form of action. The operating principle of organizations is that actors enter into and comply with commitments, and in doing so bring about the business services of the enterprise. This important insight clarifies the view that that enterprises belong to the category of social systems, i.e., its active elements (actors) are social individuals (human beings). The unifying role of human beings makes it possible to address problems in a holistic way, to achieve unity and integration in bringing about any organizational change.

Also when regarding the implementation of organizations by means of modern information technology (IT), enterprise engineering offers innovative ideas. In a similar way as the ontological model of an organization is based on atomic elements (namely, communicative acts), there is an ontological model for IT applications. Such a model is based on a small set of atomic elements, such as data elements and action elements. By constructing software in this way, the combinatorial effects (i.e., the increasing effort it takes in the course of time to bring about a particular change) in software engineering can be avoided.

The development of enterprise engineering requires the active involvement of a variety of research institutes and a tight collaboration between them. This is achieved by a continuously expanding network of universities and other institutes, called the CIAO! Network (www.ciaonetwork.org). Since 2005 this network has organized the annual CIAO! Workshop, and since 2008 its proceedings have been published as *Advances in Enterprise Engineering* in the Springer LNBIP series. From 2011 on, this workshop was replaced by the Enterprise Engineering Working Conference (EEWC). This volume contains the proceedings of the fifth EEWC, held in Prague, Czech Republic. There were 29 submissions. Each submission was reviewed by three Program Committee members and the decision was to accept 10 papers that were carefully reviewed and selected for inclusion in this volume.

The EEWC aims at addressing the challenges that modern and complex enterprises are facing in a rapidly changing world. The participants of the working conference share a belief that dealing with these challenges requires rigorous and scientific solutions, focusing on the design and engineering of enterprises. The goal of EEWC is to stimulate interaction between the different stakeholders, scientists as well as practitioners, interested in making Enterprise Engineering a reality.

June 2015

David Aveiro
Robert Pergl
Michal Valenta

Enterprise Engineering – The Manifesto

Introduction

This manifesto presents the focal topics and objectives of the emerging discipline of enterprise engineering, as it is currently theorized and developed within the CIAO! Network. There is close cooperation between the CIAO! Network (www.ciaonetwork. org) and the Enterprise Engineering Institute (www.ee-institute.com) for promoting the practical application of enterprise engineering. The manifesto comprises seven postulates, which collectively constitute the *enterprise engineering paradigm* (EEP).

Motivation

The vast majority of strategic initiatives fail, meaning that enterprises are unable to gain success from their strategy. Abundant research indicates that the key reason for strategic failures is the lack of coherence and consistency among the various components of an enterprise. At the same time, the need to operate as a unified and integrated whole is becoming increasingly important. These challenges are dominantly addressed from a functional or managerial perspective, as advocated by management and organization science. Such knowledge is necessary and sufficient for managing an enterprise, but it is inadequate for bringing about changes. To do that, one needs to take a constructional or engineering perspective. Both organizations and software systems are complex and prone to entropy. This means that in the course of time, the costs of bringing about similar changes increase in a way that is known as combinatorial explosion. Regarding (automated) information systems, this has been demonstrated; regarding organizations, it is still a conjecture. Entropy can be reduced and managed effectively through modular design based on atomic elements. The people in an enterprise are collectively responsible for the operation (including management) of the enterprise. In addition, they are collectively responsible for the evolution of the enterprise (adapting to needs for change). These responsibilities can only be borne if one has appropriate knowledge of the enterprise.

Mission

Addressing the challenges mentioned above requires a paradigm shift. It is the mission of the discipline of enterprise engineering to develop new, appropriate theories, models, methods, and other artifacts for the analysis, design, implementation, and governance of enterprises by combining (relevant parts of) management and organization science, information systems science, and computer science. The ambition is to address (all) traditional topics in said disciplines from the enterprise engineering paradigm. The result of our efforts should be theoretically rigorous and practically relevant.

Postulates

Postulate 1

In order to perform optimally and to implement changes successfully, enterprises must operate as a unified and integrated whole. *Unity* and *integration* can only be achieved through *deliberate enterprise development* (comprising design, engineering, and implementation) and *governance*.

Postulate 2

Enterprises are essentially social systems, of which the elements are human beings in their role of *social individuals*, bestowed with appropriate *authority* and bearing the corresponding *responsibility*. The *operating principle* of enterprises is that these human beings enter into and comply with *commitments* regarding the products (services) that they create (deliver). Commitments are the results of *coordination acts*, which occur in universal patterns, called *transactions*.

Note. Human beings may be supported by technical artifacts of all kinds, notably by ICT systems. Therefore, enterprises are often referred to as sociotechnical systems. However, only human beings are responsible and accountable for what the supporting technical artifacts do.

Postulate 3

There are two distinct perspectives on enterprises (as on all systems): *function* and *construction*. All other perspectives are a subdivision of one of these. Accordingly, there are two distinct kinds of models: *black-box models* and *white-box models*. White-box models are *objective*; they regard the construction of a system. Black-box models are *subjective*; they regard a function of a system. *Function is not a system property* but a relationship between the system and some stakeholder(s). Both perspectives are needed for developing enterprises.

Note. For convenience sake, we talk about the business of an enterprise when taking the function perspective of the customer, and about its *organization* when taking the construction perspective.

Postulate 4

In order to manage the complexity of a system (and to reduce and manage its entropy), one must start the constructional design of the system with its ontological model. This is a fully implementation-independent model of the *construction* and the *operation* of the system. Moreover, an ontological model has a *modular* structure and its elements are (ontologically) *atomic*. For enterprises the metamodel of such models is called *enterprise ontology*. For information systems the metamodel is called *information system ontology*.

Note. At any moment in the lifetime of a system, there is only one ontological model, capturing its actual construction, though abstracted from its implementation. The ontological model of a system is comprehensive and concise, and extremely stable.

Postulate 5

It is an *ethical necessity* for bestowing authorities on the people in an enterprise, and having them bear the corresponding responsibility, that these people are able to *internalize* the (relevant parts of the) *ontological model* of the enterprise, and to constantly validate the correspondence of the model with the operational reality.

Note. It is a duty of enterprise engineers to provide the means to the people in an enterprise to internalize its ontological model.

Postulate 6

To ensure that an enterprise operates in compliance with its *strategic concerns*, these concerns must be transformed into generic functional and constructional *normative principles*, which guide the (re-)development of the enterprise, in addition to the applicable specific requirements. A coherent, consistent, and hierarchically ordered set of such principles for a particular class of systems is called an *architecture*. The collective architectures of an enterprise are called its *enterprise architecture*.

Note. The term "architecture" is often used (also) for a model that is the outcome of a design process, during which some architecture is applied. We do not recommend this homonymous use of the word.

Postulate 7

For achieving and maintaining unity and integration in the (re-)development and operation of an enterprise, organizational measures are needed, collectively called *governance*. The *organizational competence* to take and apply these measures on a continuous basis is called *enterprise governance*.

June 2015   Jan Dietz

Organization

EEWC 2015 was the 5th Enterprise Engineering Working Conference resulting from a series of successful CIAO! Workshops and EEWC Conferences over the last years. These events were aimed at addressing the challenges that modern and complex enterprises are facing in a rapidly changing world. The participants in these events share the belief that dealing with these challenges requires rigorous and scientific solutions, focusing on the design and engineering of enterprises.

This conviction has led to the effort of annually organizing an international working conference on the topic of enterprise engineering, in order to bring together all stakeholders interested in making enterprise engineering a reality. This means that not only scientists are invited, but also practitioners. Next, it also means that the conference is aimed at active participation, discussion, and exchange of ideas in order to stimulate future cooperation among the participants. This makes EEWC a working conference contributing to the further development of enterprise engineering as a mature discipline.

The organization of EEWC 2015 and the peer review of the contributions to EEWC 2015 were accomplished by an outstanding international team of experts in the fields of enterprise engineering. The following is the organizational structure of EEWC 2015.

Advisory Board

Antonia Albani	University of St. Gallen, Switzerland
Jan Dietz	Delft University of Technology, The Netherlands
Jan Hoogervorst	Antwerp Management School, Belgium
Hans Mulder	University of Antwerp, Belgium

Conference Chairs

Robert Pergl	Czech Technical University in Prague, Czech Republic
Robert Winter	University of St. Gallen, Switzerland

Program Chairs

David Aveiro	University of Madeira, Madeira Interactive Technologies Institute and Center for Organizational Design and Engineering - INESC INOV Lisbon, Portugal
Jorge Sanz	National University of Singapore, Singapore
Jan Verelst	University of Antwerp, Belgium
Antonia Albani	University of St. Gallen, Switzerland

Organizing Chairs

Robert Pergl   Czech Technical University in Prague, Czech Republic
Michal Valenta   Czech Technical University in Prague, Czech Republic

Program Committee

Bernhard Bauer   University of Augsburg, Germany
Carlos Páscoa   Portuguese Air Force Academy, Portugal
Christian Huemer   Vienna University of Technology, Austria
Duarte Gouveia   University of Madeira, Portugal
Eduard Babkin   Higher School of Economics, Nizhny Novgorod,
                              Russia
Eric Dubois   Luxembourg Institute of Science and Technology,
                              Luxembourg
Florian Matthes   Technical University Munich, Germany
Frank Harmsen   Ernst & Young Advisory and Maastricht University,
                              The Netherlands
Geert Poels   Ghent University, Belgium
Gil Regev   École Polytechnique Fédérale de Lausanne,
                              Switzerland
Graham Mcleod   University of Cape Town, South Africa
Hans Mulder   University of Antwerp, Belgium
Henderik Proper   Luxembourg Institute of Science and Technology,
                              Luxembourg
Jan Dietz   Delft University of Technology, The Netherlands
Jan Hoogervorst   Antwerp Management School, Belgium
Jan Verelst   University of Antwerp, Belgium
Jens Gulden   University of Duisburg-Essen, Germany
João Pombinho   University of Lisbon, Portugal
Johann Eder   University of Klagenfurt, Austria
Joop Jong   Mprise, The Netherlands
José Tribolet   INESC and University of Lisbon, Portugal
Junichi Iijima   Tokyo Institute of Technology, Japan
Khaled Gaaloul   Luxembourg Institute of Science and Technology,
                              Luxembourg
Marcello Bax   Federal University of Minas Gerais, Brazil
Mauricio Almeida   Federal University of Minas Gerais, Brazil
Miguel Mira Da Silva   INESC and University of Lisbon, Portugal
Niek Pluijmert   INQA Quality Consultants, The Netherlands
Nuno Castela   Polytechnic Institute of Castelo Branco, Portugal
Olga Oshmarina   Higher School of Economics, Nizhny Novgorod,
                              Russia
Paul Johannesson   Stockholm University, Sweden
Peter Loos   University of Saarland, Germany

Philip Huysmans	University of Antwerp, Belgium
Robert Lagerström	The KTH Royal Institute of Technology, Sweden
Robert Pergl	Czech Technical University in Prague, Czech Republic
Rony Flatscher	Vienna University of Economics and Business Administration, Austria
Sanetake Nagayoshi	Waseda University, Japan
Sérgio Guerreiro	Lusófona University, Lisbon, Portugal
Steven van Kervel	Formetis, The Netherlands
Stijn Hoppenbrouwers	HAN University of Applied Sciences, The Netherlands
Sybren De Kinderen	University of Luxembourg, Luxembourg
Ulrich Frank	University of Duisburg-Essen, Germany
Ulrik Franke	Swedish Defense Research Agency, Sweden

Contents

On Enterprise Engineering and DEMO

Enterprise Operational Analysis Using DEMO and the Enterprise Operating System

Emmy Dudok[1], Sérgio Guerreiro[2], Eduard Babkin[3], Robert Pergl[4], and Steven J.H. van Kervel[5($\boxtimes$)]

[1] ProcessChemistry, Geldrop, The Netherlands
emmy.dudok@processchemistry.nl
[2] Lusófona University, Lisbon, Portugal
sergio.guerreiro@ulusofona.pt
[3] National Research University – Higher School of Economics, Moscow, Russia
eababkin@hse.ru
[4] Czech Technical University, Prague, Czech Republic
robert.pergl@fit.cvut.cz
[5] Formetis, Boxtel, The Netherlands
steven.van.kervel@formetis.nl

Abstract. Monitoring and analyzing the operation of enterprises is a key capability of Governance, Risk, and Compliance (GRC) solutions and is relevant for high-risk organizations, such as financial services. The potential of state-of-the-art process mining (data-driven process analysis) is limited by quality issues with transactional data registration and extraction. A novel approach is proposed to address these challenges: the Enterprise Operational Analysis (EOA) founded in DEMO and the Enterprise Operating System (EOS). The EOS is a software system based on enterprise engineering, and stores, interprets, and executes DEMO models as native source code. The EOS provides workflow-like capabilities and supports EOA. Combining the EOS with state-of-the-art process mining offers the following advantages: guaranteed completeness of analysis, elimination of 'mining' for events, facilitating process conformance checking, analysis on various levels of granularity from various perspectives. It enables enterprises to systematically analyze, improve and deploy business procedures. A professional business case is analyzed.

Keywords: Process mining · Enterprise operational analysis · Demo methodology · Enterprise operating system · Governance risk compliance

1 Introduction

The ongoing globalization and international trade gave rise to organizations that must comply with ever more complex regulations such as Sarbanes Oxley [32], also known as the "Public Company Accounting Reform and Investor Protection Act" and Basel III [7], a global regulatory standard on bank adequacy, stress testing and liquidity risk. Digitization of society enabled new vulnerabilities such as malware, identity theft or fraud. This applies especially to large banking conglomerates, "too large to fail", the financial instability and the global sovereign debts. A similar challenge is found in

© Springer International Publishing Switzerland 2015
D. Aveiro et al. (Eds.): EEWC 2015, LNBIP 211, pp. 3–18, 2015.
DOI: 10.1007/978-3-319-19297-0_1

high-risk organizations such as power plants and chemical refineries where an extremely rigorous compliance to regulations and policies as well as the identification, containment and mitigation of risks are of utmost importance.

These kinds of challenges in multidisciplinary research are captured by the terms "Governance, Risk management, and Compliance" (GRC). Racz et al. [29] observe that this field is still very immature and lacks well-defined shared concepts, definitions and theories. Their framework for GRC and definitions is found suitable by Verwaest [34] for well-founded scientific research and is adopted in this research.

Governance involves generic principles, guidelines, and decisions made by the board for ethical criteria, transparency, protection of reputation and proper treatment of the interests of all stakeholders. It also includes operational supervision of the way these principles, guidelines and decisions are being implemented by the management and if necessary, ad hoc adjustments are made.

Risk denotes any situation or event that may cause harm to the enterprise or any of its stakeholders. In this research we focus on risks that arise from the operation of the enterprise, i.e., human actors following business processes. Financial risks related to stock exchange, currencies etc., are out of scope. Risk involves identifying specific situations in the execution of business procedures and mitigating any consequences, at (business process) design time, runtime and real-time.

Compliance is the implementation of all externally imposed (legal) regulations in day-to-day operation. Violation of compliance exposes the enterprise to legal sanctions and claims of customers and third parties.

In this work we derive certain generic and reusable design principles for GRC. The main challenge is that the daily execution of business procedures should deliver services in such a way that GRC, efficiency and effectiveness topics are well addressed. A new approach, the Enterprise Operational Analysis (EOA), is proposed to support the engineering of enterprises that adhere to these GRC principles. EOA combines process mining with DEMO and the Enterprise Operating System (EOS). EOA provides complete transparency of the daily operation, guarantees completeness and correctness, and supports real-time monitoring and analysis.

The paper is structured as follows: Sect. 2 describes generic and reusable principles of GRC. Section 3 describes state-of-the-art process mining. Section 4 elaborates the problem definition. Section 5 describes the Enterprise Operational Analysis Approach. Section 6 describes a professional business case. Section 7 discusses conclusions and further research.

2   GRC Principles

Given the analysis results of domain-specific foundations of GRC, we derive certain generic reusable design principles of GRC, in addition to the Racz framework [29].

Principle 1: Business-Process Driven. The operation of the enterprise is fully defined by business processes. Since enterprises are complex entities, there are several important quality criteria for the way business processes are defined and specified. Most state-of-the-art BPM methodologies however are not adequate [27]. Hence, there is a need for a high-quality engineering methodology to develop and model business procedures, based on a domain ontology, that provides a complete design of an enterprise, and overcomes the many problems associated with state-of-the-art BPM modeling methods [27], further elaborated in Sect. 5.

Principle 2: Design for GRC. Engineering of business processes should meet the GRC quality criteria and provide a good degree of efficiency and effectiveness. This principle states that business processes must be well designed in a functional sense. This can be achieved only if business process models are constructed and assessed for the GRC quality criteria at design time. An empirical validation is performed using model simulation, before the system is put into operation. Shared reasoning by stakeholders is used to investigate compliance, risk conditions with mitigations and application of general governance principles. If necessary, business procedures are altered and improved, which is a specific design science cycle [22, 23]. In addition to meeting the GRC quality criteria, the daily operation must be effective and efficient. This encompasses topics such as product or service quality, customer satisfaction, production costs and resource utilization, minimizing service time and errors, transparency of production and employees.

Principle 3: Prescriptive Control. Prescriptive control of the enterprise operation compliant with the business processes is put in place. This principle states that any actor of the enterprise must obey to the business procedures, i.e., operate within the allowed state space of the business process. It must be technically impossible for any actor to deviate from the business process. This is one of the capabilities achieved by the Enterprise Operating System (EOS) [19], providing enterprise control [18] elaborated in Sect. 5.

Principle 4: Enterprise Operational Analysis. The operation of the enterprise in full production must be well monitored and analyzed. The appropriate, complete and correct monitoring and analysis of the operation of the enterprise using state-of-the-art process mining at runtime is called Enterprise Operational Analysis (EOA), see Sect. 5 expressed by this principle. With these procedures in place, it is possible to detect, predict, intervene, and prevent noncompliant behavior from taking place.

Principle 5: Enterprise Operational Control. Changing regulations, new market strategies, improved insight in business procedures or the need for any improvements require the daily operation of the organization to be adapted accordingly. This requires a redesign of the business process models, including validation and renewed deployment. This should be a recurring operation, elaborated in Sect. 5. With these capabilities the goal of operational control for organizations has been achieved. It is in fact a classic control system [15] where the organization is subjected to subsequent incremental improvements.

To address GRC, efficiency and effectiveness challenges, EOA is a mandatory capability. Without it one is operating almost in the dark, without knowing what is really happening: management cannot control, steer or improve the enterprise and the operation is prone to failure. Without it, the goal of operational control, the ongoing cycle of designing, implementing, bringing into operation, cannot be reached.

3 State-of-the-Art Process Mining

The extraction of process knowledge from transactional data as registered by corporate systems is commonly known as *process mining* [1]. The input for process mining is an *event log* that captures digital footprints on cases being executed in the process. Process mining algorithms consider activities, instances and frequencies to compute the underlying process model. Various algorithms have been developed, e.g. [5, 6, 13, 21, 35], taking different perspectives with respect to dealing with frequencies, incompleteness of data, large and real event logs, support for various workflow patterns, overfitting, underfitting, top-down or bottom-up approach, etc. Some of the latest research includes addressing mining the evolution of a drifting process [8] and reducing complexity of mined declarative process models [28, 30].

Van der Aalst et al. [3] distinguish three types of process mining: (i) *discovery* of the actual process model without prior knowledge; (ii) *conformance* of the process model and its performance with a prior known reference model; (iii) *enhancement* of a prior known process model with process knowledge. They also consider different perspectives for mining: control-flow, organization, case and time. Process mining allows for operational control by gaining impartial insight into the process execution, data-driven process improvement, compliance checking, predictive analysis, and empowering employees in taking control of their work via objective self assessment. This requires a steady connection with a process-aware information system. Process mining supports the design of such a system [2] by identifying process and GRC requirements.

Process mining is more and more applied in practice by auditors to verify compliance of a business process, process execution and governance with rules and regulations such as ISO standards and SOX legislation [32]. It allows for automatic verification of process compliance over the full range of cases instead of random sampling, guaranteeing 100 % confidence. El Kharbili et al. [14] indicate that four aspects need to be covered by business process compliance checking techniques: (i) compliance checking during the entire BPM lifecycle; (ii) compliance checking in perspectives other than the control-flow; (iii) support of visual analytics; (iv) defining and applying semantic technologies for the application of compliance checking. Three compliance perspectives are usually distinguished [31]: (i) *correct* ongoing business to ensure compliance to rules and regulations, (ii) *detect* compliancy violations in past instances and (iii) *prevent* noncompliant behavior from taking place by design. Van der Aalst and Medeiros [4] apply process mining to check for security issues in audit trails. Presence or absence of certain workflow patterns in the actual execution of the process might indicate security issues. Any non-fraudulent behavior could thereafter be supported and fraudulent behavior prevented. They suggest using control-flow simulation of the process to verify conformance to specific ordering patterns.

Process mining is rapidly gaining popularity due to a rapid growth of data, the concept and awareness of Big Data and a rapidly changing and highly competitive market [3]. As it is purely based on data, data quality is of high importance. Several quality criteria for event logs are identified [3]: event logs need to be reliable and complete, events need to be recorded based on predefined semantics and security issues need to be taken care of. Several major challenges with respect to data registration and extraction are also brought to attention [3]: data might reside in any number of IT applications, are often not registered within a process context, might contain outliers, and registration is often incomplete. Other challenges include handling complex event logs, combining with other data mining techniques, cross-organizational process mining, and improved end-user support [3].

Summarizing, process mining provides powerful tools for data-driven process analysis. However, challenges with respect to data limit its potential value.

4 Problem Definition

As evident from Sect. 3, process mining provides deep and objective insight into the operations of an organization, capturing any anomalies in ongoing and past business and providing data-driven support for process definition, monitoring and improvement. It addresses both the design for GRC and the EOA GRC principles as it supports both developing systems that comply with GRC quality criteria at design time, and monitor and analyze compliance in the daily operation of the enterprise. Process mining provides insight into current processes and allows for procedural simulation and validation of the design, eliminating any noncompliant control-flow aspects or other risks at design time. Various perspectives regarding compliance, such as control-flow, resource and data aspects can be taken into account, for example in the form of social network analysis. In addition, it also supports redesign in the context of the EOC GRC principle.

As process mining is purely based on data, challenges with respect to registration and extraction of that data greatly impact the possible application for compliance of past and ongoing business. Scattered over multiple applications, and often not registered in a process-aware manner, data can be difficult to capture and the process of converting it into the required format can have a high impact on resources. When these data registration and extraction issues can be mitigated and extraction and preparation time of data can be drastically reduced, near real-time analysis and monitoring of ongoing business becomes a viable possibility, supporting the EOA GRC principle. This allows red flagging specific procedure states at runtime, ensuring safety and mitigating operational risks.

As indicated in the business-process driven GRC principle, to support compliance of the business operation, we should be able to guarantee that the operation is completely business-process driven. Therefore, a prescriptive enforcement of the operation, or descriptive as it matches the descriptiveness of compliance models has to be put into place, in accordance with the prescriptive control GRC principle. This also mitigates data-extraction challenges in the context of process mining.

Summarizing, process mining provides support forall GRC principles. However, without effective data registration and extraction in a process-aware manner, by itself it will not attain its full potential in supporting GRC. We propose the use of DEMO as a solid foundation, to resolve these aspects. This allows process mining to be used as an effective and impartial solution to GRC. This is detailed in the next section.

5 The Enterprise Operational Analysis Approach

The proposed Enterprise Operational Approach (EOA) is founded on (i) the DEMO methodology and theories [11, 12] to develop high-quality enterprise models, (ii) the Enterprise Operating System [19], a software engine that executes DEMO models "as native source code" and (iii) state of the art process mining tools. Figure 1 depicts an overview of our approach, which allows for analyzing the enterprise's operations and designing new information systems according to the actual operation and observed GRC principles. First, the various components of Fig. 1 are described below. Then, the five GRC principles of Sect. 2 are instantiated.

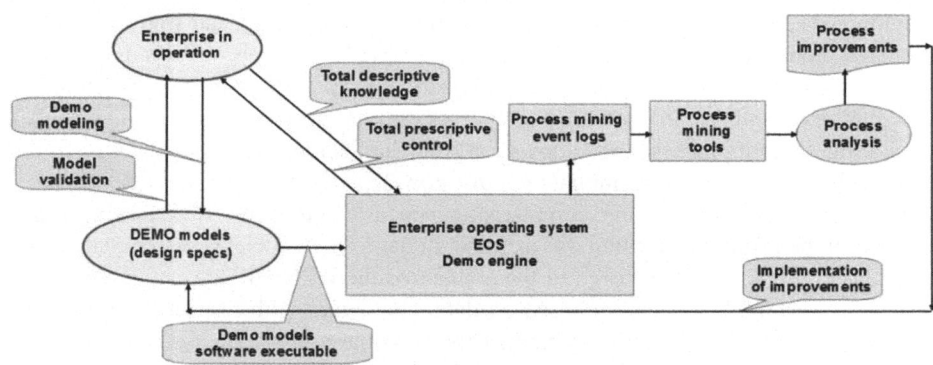

Fig. 1. Overview of the enterprise operation analysis approach

5.1 Details of the EOA Approach

Enterprise in Operation. The enterprise in operation is defined as a social system of actors who communicate about their productions [11, 12]. The system is purposefully constructed to fulfill a specific function. Actors communicate about their productions by communication acts, which result in communication facts. All communication facts represent a shared understanding and agreement of all actors about the world of production. The EOA approach is based on event logs of all communicative acts, resulting in communicative facts, of human actors about the world of productions.

DEMO Models. A great demand exists for an adequate and standard formal representation of GRC concepts [29, 33]. Taking into account strategic goals of the research and aforementioned generic principles of GRC, we provide three arguments for the direct correspondence between the conceptual structure of GRC and the foundations of DEMO modeling. First, GRC success is in part determined by the use of a proper modeling technique that reduces complexity. DEMO modeling reduces complexity due to stratification of O-, I- and D- transactions and exploits a proper level of abstraction based on a language-action-perspective. DEMO is based on an ontological theory, as defined by Enterprise Ontology [11] and is well founded on appropriate scientific theories [12]. DEMO models also provide a suitable specification of business processes

[16–18] with valuable qualities. The quality of the applied methodology is guaranteed by the underlying theories, methodologies and formal methods [9–12]. The appropriateness of DEMO is shown by business cases and applications in many domains, e.g. [24, 26]. Second, trust relations among participating business actors in the GRC domain should be explicitly determined and analyzed. The generic pattern of DEMO transactions with clear phases of communication (actagenic, action execution, factagenic) provides analysts with a powerful conceptual framework for reflection upon the trust foundations and risks between the initiator and the executor of the transaction. Finally, as the OCEG GRC Capability Model[1] determines, the key GRC activities revolve around such conceptual elements as organization boundaries, business processes, tasks, facts, policies and business rules. For each mentioned element we can easily find a direct correspondence in DEMO nomenclature. This concludes our argumentation. We can systematically apply the DEMO modeling technique for the whole process of model design and analysis activities in our approach.

DEMO Engine. The DEMO Engine is part of the Enterprise Operating System. The formal qualities of DEMO models enable the construction of this software engine that directly executes DEMO models [25] as native source code.

The Enterprise Operating System (EOS). The EOS is analogous to an operating system for a computer and represents the active layer between human actors of the organization and the enterprise information systems. The DEMO engine that executes a DEMO model constitutes the Enterprise Operating System (EOS) [19]. The EOS provides three capabilities of interest for this research: (i) *Total prescriptive control* [16–18, 20], implying that the whole enterprise, including each actor, can act exclusively within the boundaries of the (DEMO) business process. (ii) *Total descriptive knowledge.* Each communication act is captured and recorded and completeness and correctness of all acts is guaranteed. (iii) *Event Logs,* the straightforward generation of suitable event logs from recorded communication acts.

Process Mining Tools. Process mining provides data-driven process analysis and many valuable perspectives on the actual operation. For more details see Sect. 3.

Process Analysis. Process analysis refers to human actors using process mining tools to understand the operation, take appropriate actions and propose improvements for implementation, in this case improved DEMO models.

5.2   Assessment of the GRC Principles

Business-Process Driven Principle. This principle is realized by the application of DEMO modeling, providing high quality process models.

---

[1] http://thegrcbluebook.com/wp-content/uploads/2011/12/uploads_OCEG.RedBook2-BASIC.pdf.

Design for GRC Principle. DEMO models are designed by knowledgeable stakeholders using shared reasoning [11] in a design cycle [22, 23]. Process mining, model simulation and early validation are highly appropriate to design for optimal GRC support, without commitments to programming and resources [19, 25].

Prescriptive Control Principle. The EOS controls precisely which communication acts are allowed for each actor to perform. This is computed directly from the model and its current state.

EOA Principle. The EOS, which has total descriptive knowledge of the enterprise operation, allows for straightforward extraction of a guaranteed complete and correct event logs [20]. Using state-of-the-art process mining this principle is realized.

Enterprise Operational Control Principle. Combining the above principles, we realize a closed loop classical control cycle [15]. In other words, this is realized by (i) DEMO modeling; (ii) DEMO models executed by the EOS; (iii) the EOS driving the operation of the enterprise; (iv) the EOS providing complete event logs; (v) event logs processed by process mining techniques, providing data-driven process analyses that support further model improvements.

A typical challenge for process mining is that many different IT applications must be accessed to create an event log encompassing a complete business process. In the EOA approach we capture communication acts between actors about their productions [11]. It is implicitly assumed that these actors communicate in a truthful way; hence the event logs are assumed to be truthful. To verify correctness, it is recommended to cross-validate the data with various IT systems.

In general terms, EOA supports two design science engineering cycles: the modeling and model validation cycle; and the operational control cycle of model execution, logging, monitoring, analysis and implementation of improvements.

6 Case Study Representation

In this section we discuss a case study that was performed to assess the suitability of EOS as a foundation for GRC, efficiency and effectiveness checking with process mining. This case study was part of a more encompassing study, initiated by Formetis, on the general suitability of DEMO as foundation for process mining. Here we focus on the aspects of process mining relevant to GRC.

The case study was performed on data extracted from the DEMO BPM Engine of Formetis as implemented at one of their customers. It considers a process of connecting households and companies to the energy grid at a semi-public organization that delivers energy and utility services. For this case study, the process mining tool Disco[2] from software developer Fluxicon® was used.

---

[2] http://fluxicon.com/disco/.

The case study consists of several steps in which the suitability with respect to process mining is checked for: (i) the quality of data registration of the DEMO BPM Engine, (ii) the quality of data extraction from the DEMO BPM Engine and (iii) the application of process mining on data extracted from the DEMO Engine for *detective*, *corrective*, and *preventive* aspects of GRC as defined by El Kharbili et al. [14].

6.1 Transactional Data Registration

As mentioned before, process mining is fully based on transactional data, giving rise to certain challenges with respect to registration of that data. This greatly impacts the application of compliance and assurance of ongoing business. Here we evaluate to what extent the DEMO BPM Engine resolves these challenges.

As mentioned in the previous section, the DEMO BPM Engine automatically registers various atomic communication facts surrounding a specific activity or transaction performed by each individual actor. For example a request, statement of execution, and acceptance of execution of a specific transaction are registered. This way, insight can be gained in both executed transactions as well as initiated but eventually non-executed transactions. All communication surrounding a specific transaction has actually taken place, either manually or automatically. As a result, the data is highly reliable. The only remaining concern is that only communication acts are considered surrounding the actual work performed and that the actual moment of statement of work is only as reliable as the moment the resource enters it into the system. As the system is highly prescriptive, e.g. advancemend might require certain steps to be finished, it is assumed to be quite accurate.

The DEMO BPM Engine registers complete business processes, common behavior as well as exception handling, and drives several business applications. This ensures complete registration of the process within the environment of the DEMO BPM Engine. Any acts that should not be allowed are prohibited by the Engine.

Within the DEMO model, cases, transactions and communication acts are distinguished and all these entities have a predefined set of data registration attributes. This ensures high data consistency. In addition, all transactions are registered within the context of the business process as specified by the DEMO model.

6.2 Data Extraction

Similar to data registration impacting the application of compliance and assurance of ongoing business, also data extraction has to be evaluated.

Since the DEMO BPM Engine drives several business applications within the business process, data does not have to be retrieved from various applications. Instead, all data required for process mining is stored in a single central database. Desired auxiliary data residing in connected applications can also be retrieved when required. In particular, auxiliary data which is typically also used for operational management may be of interest, as they may lead to deeper analyses. However, a trade-off has to be made between required effort and impact. In this case study, we decided to use only the information readily available in the DEMO BPM Engine as implemented at the customer. This ensures generalizability to DEMO BPM Engine implementations at different entreprises.

The quality of an event log for process mining can be assessed according to a scale of maturity as described by Van der Aalst et al. [3]. Due to the complete and consistent registration and its high level of detail and reliability, event logs extracted from the DEMO BPM Engine can be ranked with 4 to 5 stars, i.e., considered to be of high quality. For this specific analysis, data was extracted from the production environment of the DEMO BPM Engine ensuring that the data has not been tempered with for testing. Additionally, data marked as sensitive to the organization (e.g. resource information) has been anonymized. See Fig. 2 for part of the event log.

```
1   Case ID,Transaction,Resource,Communication Act,Timestamp,Process Status
2   1,B-T0101,persoon14,RqAck,02/01/2013 15:46:36,1
3   1,B-T0101,rule,Pm,02/01/2013 15:46:36,3
4   1,B-T0101,persoon14,Cl,03/01/2013 10:48:08,0
5   1,B-T0101,persoon40,RqAck,08/05/2013 14:47:44,1
6   1,B-T0101,rule,Pm,08/05/2013 14:47:44,3
7   1,B-T0101,persoon40,Cl,08/05/2013 14:47:48,0
8   1,I-T0102,persoon14,StAck,02/01/2013 16:33:31,6
9   1,I-T0102,rule,Ac,02/01/2013 16:33:31,11
```

Fig. 2. Part of an event log from the DEMO BPM Engine

The data set contained a number of "legacy cases" resulting from migrating to the new system. We discarded these cases as the process execution was done only partially in this system, leading to false information about start points of the process.

6.3   Process Mining for GRC

Now that data from the DEMO BPM Engine can be considered highly suitable as input for process mining, it can be evaluated to what extent GRC is supported by this combination. The three GRC aspects of El Kharbili et al. [14] are considered: detective, corrective, and preventive.

Detective Compliance Perspective. Process mining allows for data-driven analysis of the as-is process model based on historical transactions. The process model resulting from data as registered by the DEMO BPM Engine provides a highly precise control-flow due to the various communication acts surrounding each transaction. In this respect, the actual control-flow can be compared to a reference model indicating deviations from required or agreed upon behavior. Multiple reference models were applicable in our case study, due to its time span: some performed transactions were applicable to a specific reference model and did not occur in other reference models. This allowed us to track the development of the reference models over time. Due to the high precision of data registration, various business rules with respect to control-flow can also be investigated in a highly accurate manner. Also, within the process under consideration several subprocesses were identified, allowing for a compliance check on several granularity levels within the business process (Fig. 3). Also, checks could be performed on process performance with respect to time aspects. Throughput, waiting, and processing times could be identified, again due to the accuracy with which data was available.

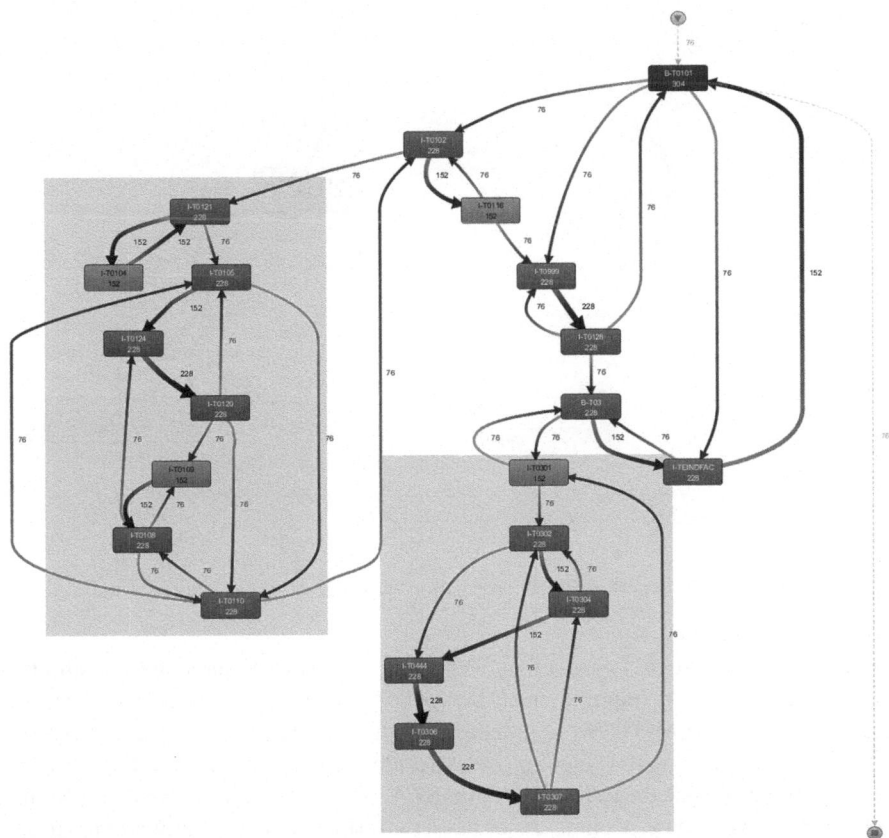

Fig. 3. Compliance checking on several granularity levels, two subprocesses are indicated

See for example Fig. 4, indicating a long waiting time on the connection and a long processing time on the right transaction. Service Level Agreements (SLAs) could be verified in either a visual way in a process model, or in various charts. One of the verified targets was a specific part of the process that had to be performed within 15 days. We found that 96 % of all cases adhered to this SLA. From an auditing point of view, each of these analyses provides a starting point from which easy drill-down and focus on anomalies is supported. Another aspect of compliance checking is the resource aspect, i.e., how resources work together. A well-known compliance check is the segregation of duties or four-eyes principle check. Such a check was performed and a total of 9 violations were found. On specific case level, more details were provided, allowing for further investigation of the root cause of the violation.

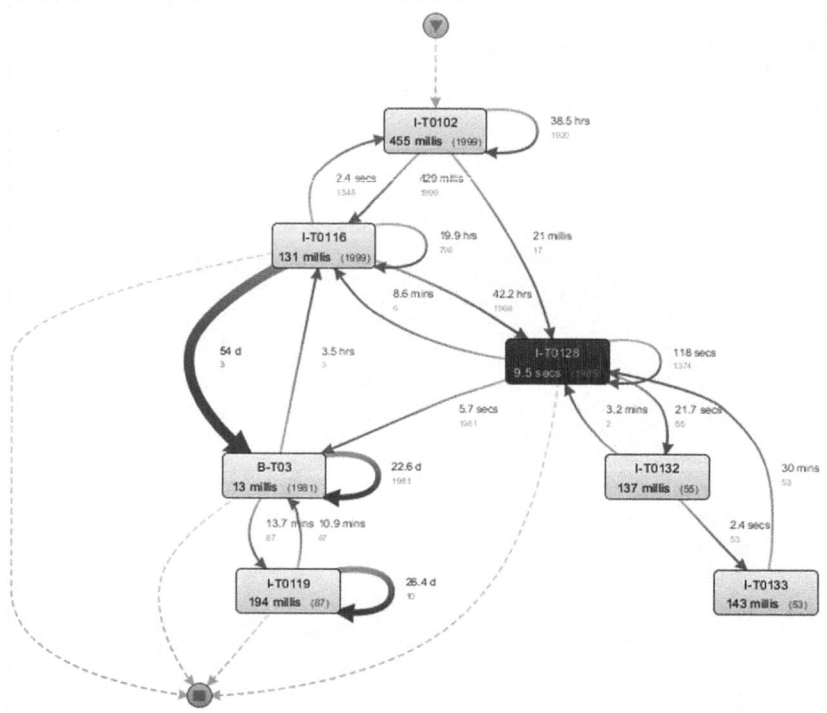

Fig. 4. Process performance information on processing and waiting times

Corrective Compliance Perspective. We have also investigated the possibility of monitoring and assuring ongoing business. As both ongoing as well as closed cases are extracted from the DEMO BPM Engine, a map could be created with process mining as to where in the process the ongoing cases reside (see Fig. 5). In addition, it can be verified whether or not they are still within SLA. This allows for corrective actions to be taken whenever a case is on a path or has a performance that is known to result in an SLA violation based on analysis on closed cases.

Preventive Compliance Perspective. Another level within the registration of this particular process in the DEMO BPM Engine is the registration of statuses in which the process can reside (see Fig. 6). We were able to identify the four most frequently occurring statuses (dark color in the figure). This view and the general process view can both be used to verify compliance with rules and regulations of the control flow already at design time based on simulation runs and analyses with process mining. Using process data extracted from the DEMO BPM Engine, process mining leads to an optimal design and a continuous fine-tuning of the WFMS during execution time to the actual behavior of the end users. Formetis is able to anticipate the desires of the end users, which are becoming transparent by analyzing their use of the software. Meaning, process mining with the DEMO Engine increases the harmonization between supplier and customers.

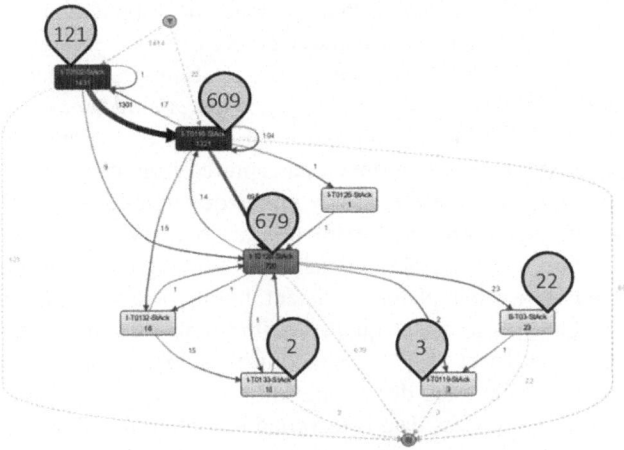

Fig. 5. Map of ongoing cases

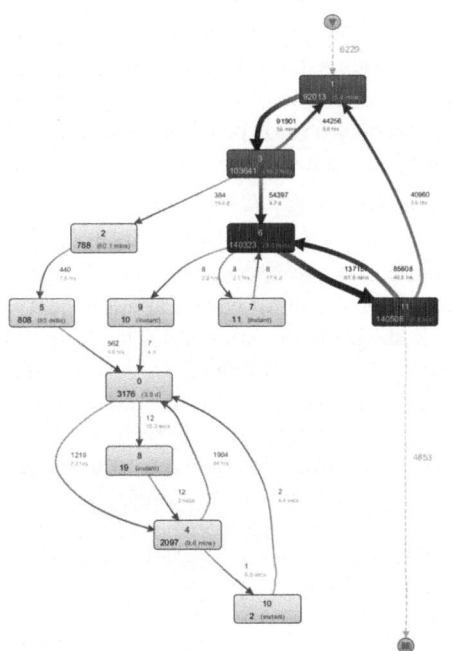

Fig. 6. Status aggregation level

7   Conclusions and Future Work

In order to address the GRC, efficiency and effectiveness challenges, enterprise operational analysis is a mandatory capability. Without it an enterprise cannot control, steer
or improve itself and is at risk of not meeting the GRC requirements. To design

effective and efficient software systems that facilitate GRC application in organizations, we proposed five GRC principles: Business process driven; Design for GRC; Prescriptive control; Enterprise Operational Analysis; Enterprise Operational Control. A novel approach is described: the *Enterprise Operational Analysis (EOA)*. EOA combines process mining with DEMO and the Enterprise Operating System (EOS). Process mining is driven by transactional data captured from corporate IT systems. As it is a data-driven technique, it relies heavily on the quality of this data. Several quality issues regarding data registration and extraction are overcome in our EOA approach. Based on a case study, we were able to reach the following conclusions. Event logs as extracted from the EOS are complete, consistent, highly detailed and reliable. As such, it is considered to be of 4- to 5-star quality on the scale of maturity of event logs as described by van der Aalst et al. [3]. Extracting these event logs from the EOS is a straightforward and efficient task, due to the full process registration occurring within the engine. Moreover, all process data is recorded in a consistent manner, in a business-process context. Finally, the advantages offered by applying process mining based on the EOS are two-fold. It enables customers to analyze, monitor and optimize their processes in a data-driven way. At the same time, it also increases the harmonization between the software supplier and their customers by providing insight into the use of their software. Moreover, combining the EOS with state-of-the-art process mining offers major advantages: guaranteed completeness of analysis, elimination of 'mining' for events, facilitating process conformance checking, analysis on various levels of granularity from various perspectives. It provides a solid foundation, enabling process mining to be used as an effective and impartial solution to GRC.

Future Work. Further research is needed on maturity of the technologies and tools, and on more empirical evidence using this approach. After all, there is only one business case investigated so far. However, the feasibility of EOA has been shown in this paper. Enterprise operational control comes at a low cost, lowering thresholds and encouraging acceptance in the professional world. The design for GRC principle can be further extended to support GRC ontologies at design time. Similarly, Sadiq and Governatori [31] propose aligning the process and control-flow aspects based on ontologies, capturing rules and regulations. Corrective GRC requires predictive analytics in process mining at real-time. Further research and development in software tools is required to further support this in the context of GRC. The proposed method aims at solving GRC, efficiency and effectiveness related issues. The fact that GRC is still considered a rather immature domain emphasizes the need for additional multidisciplinary research on GRC domain itself, and alignment of the GRC framework with the discipline of enterprise engineering.

References

1. van der Aalst, W.M.P.: Process Mining: Discovery, Conformance and Enhancement of Business Processes. Springer, Berlin (2011)
2. van der Aalst, W.M.P.: Process-aware information systems: lessons to be learned from process mining. In: Jensen, K., van der Aalst, W.M. (eds.) Transactions on Petri Nets and Other Models of Concurrency II. LNCS, vol. 5460, pp. 1–26. Springer, Heidelberg (2009)

3. van der Aalst, W., et al.: Process mining manifesto. In: Daniel, F., Barkaoui, K., Dustdar, S. (eds.) BPM Workshops 2011, Part I. LNBIP, vol. 99, pp. 169–194. Springer, Heidelberg (2012)
4. van der Aalst, W.M.P., de Medeiros, A.K.A.: Process mining and security: detecting anomalous process executions and checking process conformance. In: Electronic Notes in Theoretical Computer Science, vol. 121, pp. 3–21 (2005)
5. van der Aalst, W.M.P., de Medeiros, A.K.A., Weijters, A.: Genetic process mining. In: Ciardo, G., Darondeau, P. (eds.) ICATPN 2005. LNCS, vol. 3536, pp. 48–69. Springer, Heidelberg (2005)
6. van der Aalst, W.M.P., Weijters, A.J.M.M., Maruster, L.: Workflow mining: discovering process models from event logs. IEEE Trans. Knowl. Data Eng. **16**, 1128–1142 (2003)
7. Basel Committee on Banking Supervision in 2010–11
8. Bose, R.P.J.C., van der Aalst, W.M.P., Zliobaite, I., Pechenizkiy, M.: Dealing with concept drifts in process mining. IEEE Trans. Neural Netw. Learn. Syst. **25**(1), 154–171 (2014)
9. Ciao! Consortium; Cooperation & Interoperability - Architecture & Ontology. www.ciaonetwork.org
10. DEMO Knowledge Centre, Design and Engineering Methodology for Organizations (2012). www.demo.nl
11. Dietz, J.L.G.: Enterprise Ontology: Theory and Methodology. Springer, Heidelberg (2006)
12. Dietz, J.L.G., Hoogervorst, J.A.P.: The discipline of enterprise engineering. Int. J. Organ. Des. Eng. **3**(1), 86–114 (2013)
13. van Dongen, B.F., van der Aalst, W.M.: Multi-phase process mining: building instance graphs. In: Atzeni, P., Chu, W., Lu, H., Zhou, S., Ling, T.-W. (eds.) ER 2004. LNCS, vol. 3288, pp. 362–376. Springer, Heidelberg (2004)
14. El Kharbili, M., de Medeiros, A.K.A., Stein, S., van der Aalst, W.M.P.: Business process compliance checking: current state and future challenges. In: MobIS, vol. 141, pp. 107–113 (2008)
15. Franklin, F., Powell, D., Emami, A.: Feedback Control of Dynamic Systems, 6th edition. Addison-Wesley, Massachusetts (2009)
16. Guerreiro, S.: Enterprise governance enforcement in the operation of the runtime transactions using DEMO and ACM. In: Enterprise Engineering Working Conference (2011)
17. Guerreiro, S., Vasconcelos, A., Tribolet, J.: Adaptive access control modes enforcement in organizations. In: Quintela Varajão, J.E., Cruz-Cunha, M.M., Putnik, G.D., Trigo, A. (eds.) CENTERIS 2010. CCIS, vol. 110, pp. 283–294. Springer, Heidelberg (2010)
18. Guerreiro, S., van Kervel, S.J., Vasconcelos, A., Tribolet, J.: Executing enterprise dynamic systems control with the demo processor: the business transactions transition space validation. In: Rahman, H., Mesquita, A., Ramos, I., Pernici, B. (eds.) MCIS 2012. LNBIP, vol. 129, pp. 97–112. Springer, Heidelberg (2012)
19. Guerreiro, S., Kervel, S., Babkin, E.: Towards devising an architectural framework for enterprise operating systems. In: Proceedings of 8th International Conference on Software Paradigm Trends (2013)
20. Guerreiro, S., Tribolet, J.: Conceptualizing enterprise dynamic systems control for run-time business transactions. In: Proceedings of 21st European Conference on Information Systems, paper 56 (2013)
21. Günther, C.W., van der Aalst, W.M.: Fuzzy mining – adaptive process simplification based on multi-perspective metrics. In: Alonso, G., Dadam, P., Rosemann, M. (eds.) BPM 2007. LNCS, vol. 4714, pp. 328–343. Springer, Heidelberg (2007)
22. Hevner, A.R., March, S.T., Park, J., Ram, S.: Design science in information systems research. MIS Q. **28**(1), 75–105 (2004)

23. Hevner, A.R.: A three cycle view of design science research. Inf. Syst. Decis. Sci. Scandinavian J. Inf. Syst. **19**(2), 87–92 (2007)
24. Hintzen, J., van Kervel, S.J.H., van Meeuwen, T., Vermolen, J.A.J., Zijlstra, B.: A professional case management system in production, modeled and implemented using DEMO. In: Proceedings of 16th IEEE Conference on Business Informatics (2014)
25. van Kervel, S.J.H., Dietz, J.L.G., Hintzen, J., van Meeuwen, T., Zijlstra, B.: Enterprise ontology driven software engineering. In: Proceedings of 7th International Conference on Software Paradigm Trends (2012)
26. van Kervel, S.J.: High quality technical documentation for large industrial plants using an enterprise engineering and conceptual modeling based software solution. In: De Troyer, O., Bauzer Medeiros, C., Billen, R., Hallot, P., Simitsis, A., Van Mingroot, H. (eds.) ER Workshops 2011. LNCS, vol. 6999, pp. 383–388. Springer, Heidelberg (2011)
27. Van Nuffel, D., Mulder, H., Van Kervel, S.: Enhancing the formal foundations of BPMN by enterprise ontology. In: Albani, A., Barjis, J., Dietz, J.L. (eds.) CIAO! 2009. LNBIP, vol. 34, pp. 115–129. Springer, Heidelberg (2009)
28. Pesic, M., van der Aalst, W.M.: A declarative approach for flexible business processes management. In: Eder, J., Dustdar, S. (eds.) BPM Workshops 2006. LNCS, vol. 4103, pp. 169–180. Springer, Heidelberg (2006)
29. Racz, N., Weippl, E., Seufert, A.: A frame of reference for research of integrated governance, risk and compliance (GRC). In: De Decker, B., Schaumüller-Bichl, I. (eds.) CMS 2010. LNCS, vol. 6109, pp. 106–117. Springer, Heidelberg (2010)
30. Richetti, P.H., Baião, F.A., Santoro, F.M.: Declarative process mining: reducing discovered models complexity by pre-processing event logs. In: Sadiq, S., Soffer, P., Völzer, H. (eds.) BPM 2014. LNCS, vol. 8659, pp. 400–407. Springer, Heidelberg (2014)
31. Sadiq, S., Governatori, G.: A methodological framework for aligning business processes and regulatory compliance. In: vom Brocke, J., Rosemann, M. (eds.) Handbook of BPM, vol. 2, pp. 159–176. Springer, Heidelberg (2009)
32. Sarbanes–Oxley Act of 2002 (2002). Sarbanes-Oxley Act. Pub.L. 107–204, 116 Stat. 745, enacted 30 July 2002
33. Spies, M., Tabet, S.: Emerging standards and protocols for governance risk and compliance management. In: Kajan, E., Dorloff, F.D., Bedini, I. (eds.) Handbook of Research on E-Business Standards and Protocols: Documents, Data, and Advanced Web Technologies. IGI Global, Hershey (2012)
34. Verwaest, J.: Towards a comprehensive methodology for the implementation of Enterprise Governance & Assurance of IT, based on COBIT5, using the discipline of enterprise engineering. MSc. thesis, Antwerp Management School (2013)
35. Weijters, A.J.M.M., van der Aalst, W.M.P., de Medeiros, A.K.A.: Process mining with the heuristics miner-algorithm. In: BETA Working Paper Series, WP 166 (2006)

Engineering the Decision-Making Process Using Multiple Markov Theories and DEMO

Sérgio Guerreiro(✉)

Lusófona University, Campo Grande 376, 1749-024 Lisbon, Portugal
sergio.guerreiro@ulusofona.pt

Abstract. In the current fast-changing and turbulent operational environments, the organizations are continually being pressured by many endogenous and exogenous environmental variables. Many and complex effects occur simultaneously and large volumes of data are available. For this reason, in a process-based organization, when change is demanded (*e.g.*, business processes re-engineering) it is difficult to collect, and interpret, the complete information about the current state of the organization. Therefore, a problem is how to decide which design actions should be enacted with the incomplete information available from the executed business processes. In this context, this paper combines information systems engineering (DEMO business transactions design) and operation research (Markov theories) to contribute to the decision-making body of knowledge. As the result, this solution enforces the organization with resiliency capabilities that are triggered whenever any misalignment occurs. The proposed solution is evaluated through argumentation and by a qualitative comparison between two Markov theories (MDP and POMDP) based on a real-world case study.

Keywords: Decision-making · Management · MDP · Observation · POMDP · State · Value

1 Introduction

Decision-making is a management competence [16,20] that encompasses: the intelligence to discover the organizational problems, the design of potential solutions, the choosing of the best solution, the implementation of the solution and the verification if the new solution fulfills the desired goals. These stages occur in many levels of organizational management, *e.g.*, project management, operational management, middle management, *etc.*.

On the one hand, multiple endogenous and exogenous factors promote the need to enforce a continuous decision-making process, for instance, requirements change, legal changes or fraud attempts. In response to these multiple changes, it is necessary to have native decision-making capabilities that continuously find innovative solutions to adapt the organizational operation to be more efficient

© Springer International Publishing Switzerland 2015
D. Aveiro et al. (Eds.): EEWC 2015, LNBIP 211, pp. 19–33, 2015.
DOI: 10.1007/978-3-319-19297-0_2

and effective. In this context, the study of mechanisms to engineer the informed decision-making [24] are key competence for the success of the organization's management.

On the other hand, the business processes play a dual role: *(i)* they are the result of applying design constraints for a particular organizational reality [15], and are valid over a given period of time, and *(ii)* operational support to the actions performed by actors, by other words, business process guide actors in acting. The actors have an active and autonomous role in the execution of business processes, therefore, it is not guaranteed that the requirements of business processes are met properly on the daily routines. For example, if a company's recommendation is to always obtain a written record when contacts are made to the clients, nothing limits the ability of an actor to contact a client directly, by phone, without leaving any trace of the communication. The same example can be applied to the financial markets, with a huge adverse impact potential to the organization and to its environment.

In this context, combining decision-making with business processes, under complex process-based environments, raise the following challenges *(i)* inability to map the current operational observations with the current state where the organization actually is [17], *e.g.*, when actors perform workarounds [1] and override the previous defined prescriptions then the manager need to collect more information to interpret what, in fact, was executed; and *(ii)* incomplete observations [3], *e.g.*, because its too expensive to collect information, or, if the business processes are partially performed in paper by humans and partially machine-based. Therefore, in the majority of the situations, the management should support their decisions in partial information about the surrounding environment (also named as partial observable environments).

In light of this, in this paper, we narrow the decision-making management problem to the business transactions operation optimization. So forth, we propose and evaluate an innovative approach combining DEMO-based business process design [5,7]) and operations research (using Markov decision process (MDP) and partial observable Markov decision process (POMDP) theories [19]). DEMO obliges the full specification of business transaction dynamics and MDP and POMDP yields the greatest amount of utility over some number of decision steps.

Figure 1 provides an overview of our approach. The steering cycle of observation (cf. Fig. 1(1)), assessing the environment (cf. Fig. 1(2)), designing the potential solutions (cf. Fig. 1(3)) and choosing the best solution (cf. Fig. 1(4)). These steps recall to the management competences and we emphasize that they are mainly human based. Nevertheless, we argue and show how automatic tools deliver support to the managers, aiding at some point in their decision-making tasks.

The rest of the paper follows a simplified design science research (DSR) approach [14], encompassing the iterations of: problem statement, design of a solution for the given problem and evaluation phase. Firstly, Sect. 2 identifies the problem statement boundary and the background concepts (MDP, POMDP and DEMO) that are needed in the rest of the paper. Then, in Sect. 3, the design

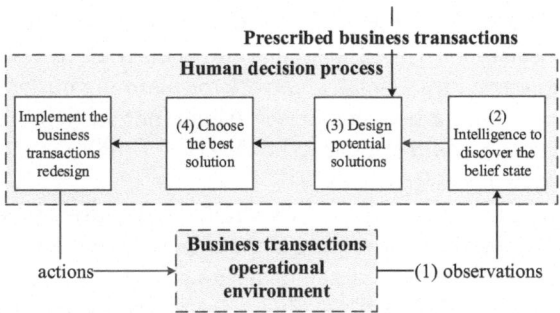

Fig. 1. Overview of our approach.

for an informed decision-making process is detailed. After that, Sect. 4 is devoted to the explanation of a case study that was previously introduced in [11]. Next, from the preceding results, Sect. 5 evaluates the solution for the given case study, using argumentation and a qualitative comparison between MDP and POMDP when applied to the context of business transactions. Afterwards, Sect. 6 presents the conclusions and future work.

2 Background Concepts

2.1 Markov Theories

In probabilities theory, a Markov process is a stochastic process that satisfies the Markov property [19]: if the transition probabilities from any given state depend only on the actual state and not on previous history. By other words, the predictions for the future are solely based on its present state. Its future and past are independent. Markov theories are applied to systems that are controlled or uncontrolled (autonomous) *versus* observable or partial observable. Where, a system is completely observable if every state variable of the system affects some of the outputs. And, a process is said to be completely controllable if every state variable of the process can be controlled to reach a certain objective in finite time by some unconstrained control action.

A Markov chain is used to refer to a process which has a countable and discrete set of state spaces, yet not controllable. When the states of the process are only partial observable, then an hidden Markov model (HMM) should be used. From this point forward, to engineer the decision-making process, we narrow our research in the controllable systems.

A Markov decision process (MDP) is able to solve the problem of calculating an optimal policy in an accessible and stochastic environment with a known transition model [18]. A MDP is defined by the tuple (S, A, T, R, γ).

In partial accessible environments, or whenever the observation does not provide enough information to determine the states or the associated transition probabilities, then the hidden Markov model (HMM) or partially observable Markov decision process (POMDP) solutions should be considered. The difference is that HMM is applied to uncontrolled systems and POMDP to controlled systems. A POMDP solution provides a rich framework for planning under uncertainty [11]. A POMDP finds a mapping between observations (not states) to actions. In practice, two different states could appear to be observed equally. A POMDP is defined by the tuple $(S, A, Z, T, O, R, \gamma)$.

The definitions for the MDP and POMDP tuples are: S is a set of states, representing all the possible underlying states the process can be in, even if state is not directly observable; A is a set of actions, representing all the available control choices at each point in time; Z is a set of observations, consisting of all possible observations that the process can emit; $T : A \times S \times S \rightarrow \prod(S)$ is a state transition function, where $\prod(.)$ is a probability distribution over some finite set, encoding the uncertainty in the system state evolution; $O : A \times S \times Z \rightarrow \prod(Z)$ is an observation function, relating the observations to the underlying state; $R : A \times S \times S \times Z \rightarrow \mathbf{R}$ is an immediate reward function, giving the immediate utility for performing an action of the underlying states; γ: discounted factor of future rewards, meaning the decay that a given achieved state suffers through out time.

For a POMDP, at each period, the environment is in some state $s \in S$. The manager takes an action $a \in A$, which causes the environment to transition to state S' with probability $T(S'|S, a)$. And because the manager does not know the exact state the system is then the manager must estimate a probability distribution, known as *belief state*, over the possible states S. This estimation is used as a seed to be refined by the POMDP executions.

Figure 2(1) presents a system transiting from state S to state S', supported by MDP. Also, Fig. 2(2) presents a diagram with a system transiting from state S to state S', supported in a partial observation, and using a belief state to achieve the reward on S'. Without knowing the actual state S at time t (cf. Fig. 2(2)), the partial observation triggers the possibility of having one or more belief states. The challenge of solving a POMDP is to maximize the reward of a given action A achieving the state S' at time $t + 1$, from the belief states. In the end, a control policy will yield the greatest amount of utility over some number of decision steps. As a summary, both POMDP and MDP require a set of states, a set of actions, transitions and rewards. The actions' effects on the state in a POMDP is exactly the same as in an MDP. The difference is in whether or not we can observe the current state of the process. In a POMDP we add a set of observations to the model. So instead of directly observing the current state, we obtain an observation which provides a hint about what state it is in. The observations are probabilistic; therefore, an observation model encompassing the probability of each observation for each state in the model should be defined.

2.2 DEMO Theory and Methodology

From the business processes point of view, DEMO theory and methodology [5] introduces capabilities to deal rigorously with the dynamic aspects of the process-based business transactions using an essential ontology that is compatible with

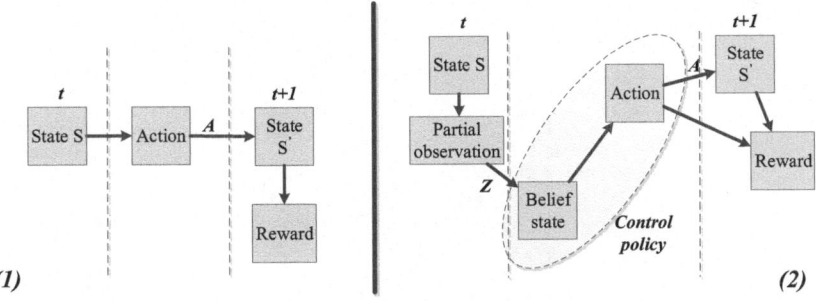

Fig. 2. State transition from state S to state S'. MDP (1) and POMDP (2) solutions.

the communication and production, acts and facts that occur between actors in the different layers of the organization. A DEMO business transaction model [6] encompasses two distinct worlds: *(i)* the transition space and *(ii)* the state space.

On the one hand, the DEMO transition space is grounded in a theory named as Ψ-theory (PSI), where the basic transaction pattern includes two distinct **actor roles**: the Customer and the Producer. The goal of performing such a transaction pattern is to obtain a new fact. The transactional pattern is performed by a sequence of coordination and production **acts** that leads to the production of the new fact. In detail, encompasses: *(i)* order phase that involves the acts of request, promise, decline and quit, *(ii)* execution phase that includes the production act of the new fact itself and *(iii)* result phase that includes the acts of state, reject, stop and accept. Firstly, when a Customer desires a new product, he requests it. After the request for the production, a promise to produce the production is delivered by the Producer. Then, after the production, the Producer states that the production is available. Finally, the Customer accepts the new fact produced. DEMO basic transaction pattern aims specifying the transition space of a system that is given by the set of allowable sequences of transitions. Every state transition is exclusively dependent from the current states of all surrounding transactions. There is no memory of previous states. This memoryless property holds with Markov theories. On the other hand, the DEMO state space delivers the model for the business transactions **facts**, which are products or services, and that are obtained by the business transaction successful execution. Throughout the business transaction execution more intermediate facts are required.

Based in the stated above, we conceptualize the DEMO business transactions as a set of triples using the dimensions of: **actor roles**, **acts** and **facts**. This conceptualization could also be aligned with the Subject-oriented Business Process Management [8] work where the three core dimensions of a business processes are: **subject**, **predicate** and **object**. This possible alignment will be further assessed in detail in the near future.

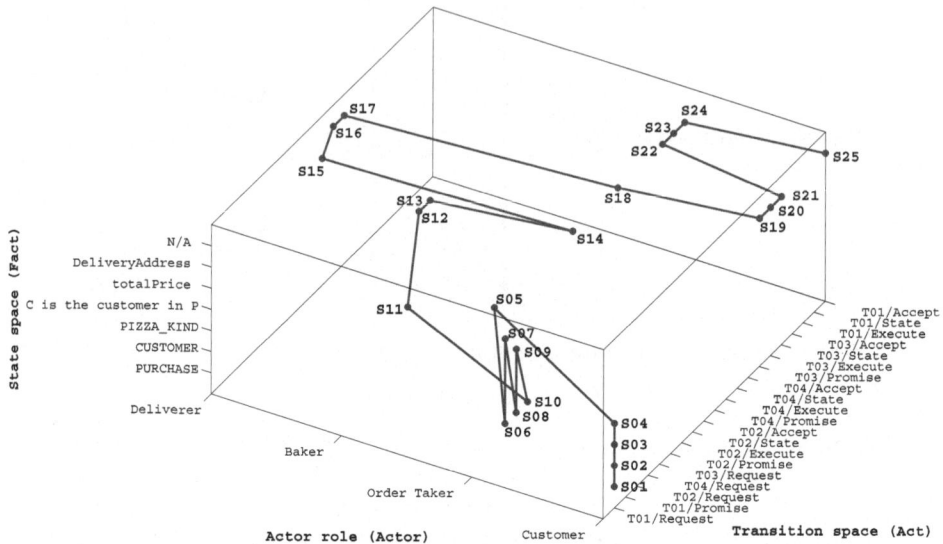

Fig. 3. Business transaction space prescription example. Adapted from [13]

3 Solution

First, we anchor the DEMO business transaction model definition in the partial observable Markov decision process (POMDP) specification. For that end, each DEMO business transaction (BT) state is defined by the following triple: $S_i =<Act_i, Fact_i, ActorRole_i >$, whereas, Act represents coordination or production act of a BT, $Fact$ represents a fact related with a BT, $ActorRole$ represents the performing actor role involved in the Act or Fact, and i identifies the state. The set of triples S represent the trajectory prescribed for the organization. Figure 3 exemplifies one possible trajectory prescription[1] as previously introduced in [13].

However, considering a partial observable system, each observation cannot be directly related with a specific S_i. By other words, what we observe from a state is not the same as the state itself. For instance, the state of order deliver could be partially observed by a signed document by the customer. Therefore, the order deliver state is an abstraction, that by its turn, is instantiated in the operation of the organization, when a document is signed. Leaving to the manager, the difficult task of relating a signed document with the achievement of a given state. In this sense, our solution, follows the POMDP premise that an observation does not correspond to a 1:1 state definition. In POMDP, each observation is used to compute the state where the system is believed to be (belief state).

The pseudo-code of the informed decision-making solution is given in Algorithm 1. The method starts by modeling the business transactions using a

---

[1] These S triples conform to the representation proposed by [23] where a triple describes each system state and supports the subsequent simulation results.

methodology with (at least) the capabilities of modeling the transition, the state and actor role spaces, *e.g.*, DEMO [5], SBPM [8], BPMN [10], *etc.* Afterwards, the business transactions models are converted in a set of memory less triples as introduced previous in this section (cf. Fig. 3). The advantage of decoupling steps 1 and 2 is because is easier to find the triples after producing a business transactions model. Then, step 3, P is populated with the POMDP tuple estimation. Usually, it corresponds to a file creation in the POMDP format[2]. To facilitate the generation of the POMDP file (summing up to 7500 configuration lines in our case study, cf. Sect. 4) a JAVA application was specially developed for this purpose. After that, from step 4 until 11, the POMDP is computed: (1) execute the action that the current node tells us; (2) receive the resulting observation from the world; (3) transition to next node based on the observation; (4) repeat to step (1). In the end, a policy graph mapping $Z \to A$ is delivered (cf. shadowed ellipse area in Fig. 2). Finally, the policy graph is rendered using any graphical tool.

Algorithm 1. Method to compute the informed decision-making.

Require: Business transaction prescriptions
Ensure: Control policy graph ($Z \to A$)
 1: **Set** $M \leftarrow$ Model the prescribed business transactions.
 2: Convert M in a set of triples: $S_i =< Act_i, Fact_i, ActorRole_i >$.
 3: $P \leftarrow$ POMDP tuple $(S, A, Z, T, O, R, \gamma)$ estimation.
 4: **for all** node of P **do**
 5:     **for each** Z **do**
 6:         Calculate $Prob(Z)$
 7:         Calculate Belief State
 8:         Calculate **R**
 9:         Calculate A
10:     **end for**
11: **end for**
12: Render the computed policy graph ($Z \to A$) using a graphical tool.

---

4   Case Study

An agro-food industry company focusing the transformation of fresh fruits to preparations that are sold to other companies is considered. Its clients are industries of milk-based products, ice creams, cakes and beverages products. To guarantee the product quality, fruit producers are subject to a ratification process before starting supplying fruit. The fruit passes through three stages: *(i)* raw material, *(ii)* ingredients after raw material preparation, and *(iii)* finished product after ingredients transformation. Until reaching the end consumer, a complex value chain is executed including the actor roles of: client, fruit producer, raw material receptionist, ingredient preparator (*e.g.*, weighing and cleaning),

---

[2] An example of this standard format could be consulted at [2].

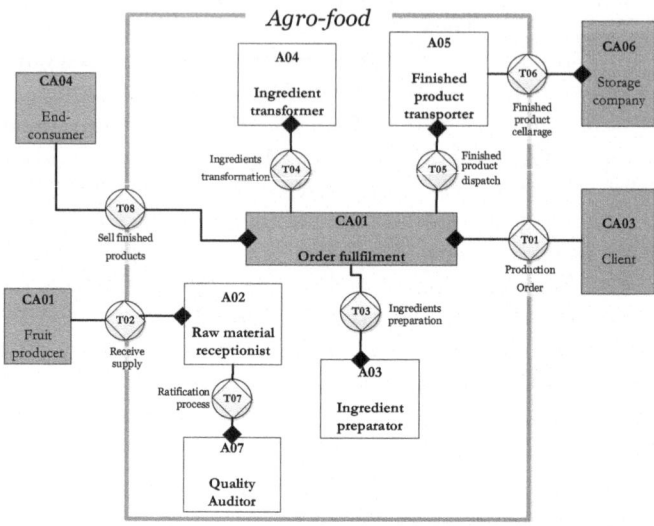

Fig. 4. Ator transaction diagram of Agro-food case study.

ingredient transformer (*e.g.*, mixing components, adding water, sugar or other products accordingly with the recipe), finished product transporter and storage company (when the agro-food company is not able to locally store all the production). The production starts when a client order is received (produce to order policy). Then, five stages are performed: receive supply, ingredients preparation, ingredients transformation, finished product cellarage and dispatch. Besides selling to other companies, they also sell a small part of finished products directly to the end consumer. Table 1 explains the result obtained with each business transaction. In more detail, Fig. 4 depicts the business transactions involving the actor roles by an actor transaction diagram (ATD) in DEMO [5].

Due to the value chain raising complexity, including many other companies (*e.g.*, suppliers), and also, due to the food safety legal obligations, traceability is a core functionality to identify the products throughout the production value chain. It encompasses three basic considerations: the product identification, the product origin and the product destination. When a lot infection is detected, traceability aids the identification of its location and removing it from the market. A lot infection may occur due to many workarounds, *e.g.*, recipe not followed by ingredient transformer, allergenic material infection at ingredient preparation, contamination during transportation, bad temperature conditions for transportation or fruit disruption stock.

Table 2 synthesize the POMDP variables that are estimated for this case study[3]. To begin with, S is given by the ATD DEMO model depicted in Fig. 4,

---

[3] The full POMDP file is public available with doi:10.13140/2.1.4433.2326.

Table 1. Transaction product table of Agro-food case study.

Transaction kind	Product kind
T01 - Production order	P01 - Client Order CO is completed
T02 - Receive supply	P02 - Supply Order SO is completed
T03 - Ingredients preparation	P03 - Ingredients I of Client Order are prepared
T04 - Ingredients transformation	P04 - Ingredients I of Client Order are transformed
T05 - Finished product dispatch	P05 - Finished product FP of Client Order is dispatched
T06 - Finished product cellarage	P06 - Finished product FP of Client Order is stored
T07 - Ratification process	P07 - Process P is ratified
T08 - Sell finished products	P08 - Finished product FP is sold

and in detail by the DEMO business transaction space (cf. Fig. 5) where each state is grounded by the triple $S_i = < Act_i, Fact_i, ActorRole_i >$. For clear explanation, the triple is simplified by the pair: $S_i = < Act_i, ActorRole_i >$ avoiding the fact types involved in all business transactions being. A full usage of DEMO business transaction space is explained in [12]. Nevertheless, this simplification does not affect the nature of the obtained decision-making results and consequent conclusions. In the right part of Fig. 5, four flows of work operated by the organization are identified: *(i)* production to client order, *(ii)* cellarage, *(iii)* fruit supply and *(iv)* selling products to end consumer.

Recalling Table 2, Z and A represent respectively the observation and the actions. Z differs from S in the sense that is much simpler and is totally business-oriented. Managers are able to observe if products are being delivered correctly, if any complaint was received, if stock is below a certain threshold, if any problem occurred while the products are being transported or else if it is running correctly (OK) so far. Therefore, in this example, there exist 40 possible states defined from the business transactions model (cf. Fig. 5), but only 5 possible observations may occur in operation.

The POMDP variable A specifies the capability of management to take actions. Four distinct actions are possible: no action (*no_op*), to cancel a previous order requested by a client (*cancel_client_order*), to request more ingredients from a fruit producer (*request_more_ingredients*) and to send quality questionnaire to the clients to assure their level of satisfaction (*send_quality_questionnaire*).

Regarding T matrix, for each action a probability is estimated assuming that transition from an initial to a final state occurs. On the one hand, the *no_op* action do not have impact in the normal progress of the business transactions operation. For simulation purposes, we assumed that in 95 % of situations if *no_op* is enacted then the states follows as described in Fig. 5. However, when action *send_quality_questionnaire* is enacted then a 50 % chance of sending it to the client exists if the selling business transactions are at stating transition step (S19 and S39). On the other hand, *cancel_client_order* and *request_more_ingredients* actions changes their normal operational progress. The first one, restarts the produce to order flow (jump to S01), and the second one, invokes the fruit supply flow (jump to S26). With correspondingly 90 % and 75 % chance of happening.

Table 2. POMDP variables definition for Agro-food case study.

POMDP variable	Value
S	40 states (S01 ... S40) following the DEMO standard pattern of a transaction: { $<$ $T01\_request, Client>, \ldots, < T01\_accept, Client>, \ldots, < T08\_request, End\ consumer>, \ldots, < T08\_accept, End\ consumer>$ }. Full specification in Fig. 5
Z	{$product\_delivered,\ complaint,\ stock\_break,\ transport\_disrupt,\ running\text{-}ok$}
A	{$no\_op,\ cancel\_client\_order,\ request\_more\_ingredients,\ send\_quality\_questionnaire$}
$T : A \times S \times$ $S \to \prod(S)$	$A = no\_op \to$ Proceed states cf. Fig. 5 by 95 % of situations
	$A = send\_quality\_questionnaire$ AND S $\subset [19, 39] \to$ true in 50 % of situations
	$A = cancel\_client\_order$ AND S $\subset [1..20] \to$ restart S01 **else** proceed states cf. Fig. 5 by 90 % of situations
	$A = request\_more\_ingredients$ AND S $\subset [1..20] \to$ invoke fruit supply (S26) **else** proceed states cf. Fig. 5 by 75 % of situations
$O : A \times S \times$ $Z \to \prod(Z)$	Regarding the end state of producing and selling products:
	$S = (20\ OR\ 40)$ AND $Z = product\_delivered \to 70\,\%$
	$S = (20\ OR\ 40)$ AND $Z = stock\_break \to 10\,\%$
	$S = (20\ OR\ 40)$ AND $Z = complaint \to 10\,\%$
	$S = (20\ OR\ 40)$ AND $Z = (transport\_disrupt\ OR\ running\_ok) \to 5\,\%$
	Regarding cellarage end state: $S = 23$ AND $Z = transport\_disrupt \to 20\,\%$ **else** 80 %
	Regarding all other end states: $Z = running\_ok \to 95\,\%$ **else** 5 %
$R : A \times S \times$ $S \times Z \to \mathbf{R}$	Flow of work ends: {$S20, S25, S35, S40$} $\to \mathbf{R} = 1$
	$complaint\ OR\ stock\_break\ OR\ transport\_disrupt$ are observed $\to \mathbf{R} = -5$
γ	5 %
Start state	$S01$
Start action	$request\_more\_ingredients$

Regarding the O matrix, for all A that moves to end state S it delivers an observation Z with probability P. The estimation follows the reasoning: in the majority of the situations (95 %) *running_ok* is observed. When the end state of producing (S20) or selling the products (S40) is achieved, the observations of *stock_break* or *complaint* could happen with 10 % probability each. Also, when the cellarage is being executed (S23) the *transport_disrupt* could be observed with 20 % probability, *e.g.*, when a truck has an accident.

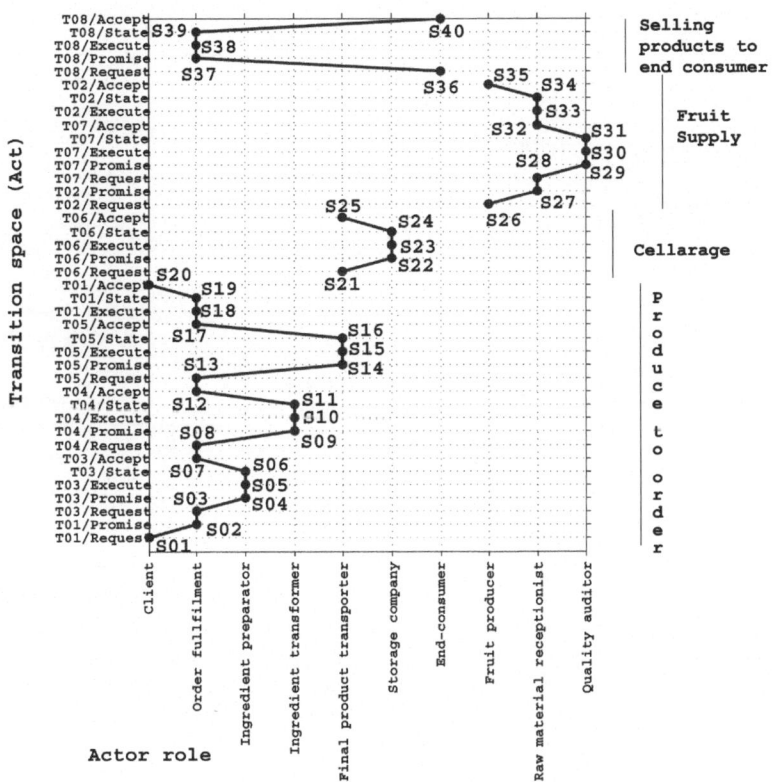

Fig. 5. Model of business transaction space for Agro-food case study (40 states). Left axis: acts, right axis: flows of work, and bottom axis: actor roles.

Finally, regarding R matrix, if any flow of work is terminated successfully then a reward of 1 is assigned. Otherwise, a negative reward of -5 is used as a penalty.

5   Qualitative Evaluation

We operationalized our proposal for the Agro-food case study by applying MDP and POMDP solvers. The POMDP was performed cf. Algorithm 1 where the solver is APPL toolkit [2], which is a recent C++ implementation running in Linux environment[4]. The delivered policy graph is rendered using GraphViz [9] tool. The MDP is computed by a $Matlab^{©}$ toolbox[5] using a linear programming algorithm. The intent of our proposal is to explore the benefits of using stochastic approaches to aid the management decisions. This goal can be achieved if

---

[4] Others POMDP solvers are available, *e.g.*, Perseus [22] implementation of randomized point-based approximate value, Tony Cassandra [4] solver, the ZMDP solver for POMDP and MDP [21].

[5] Toolbox public available at http://www7.inra.fr/mia/T/MDPtoolbox.

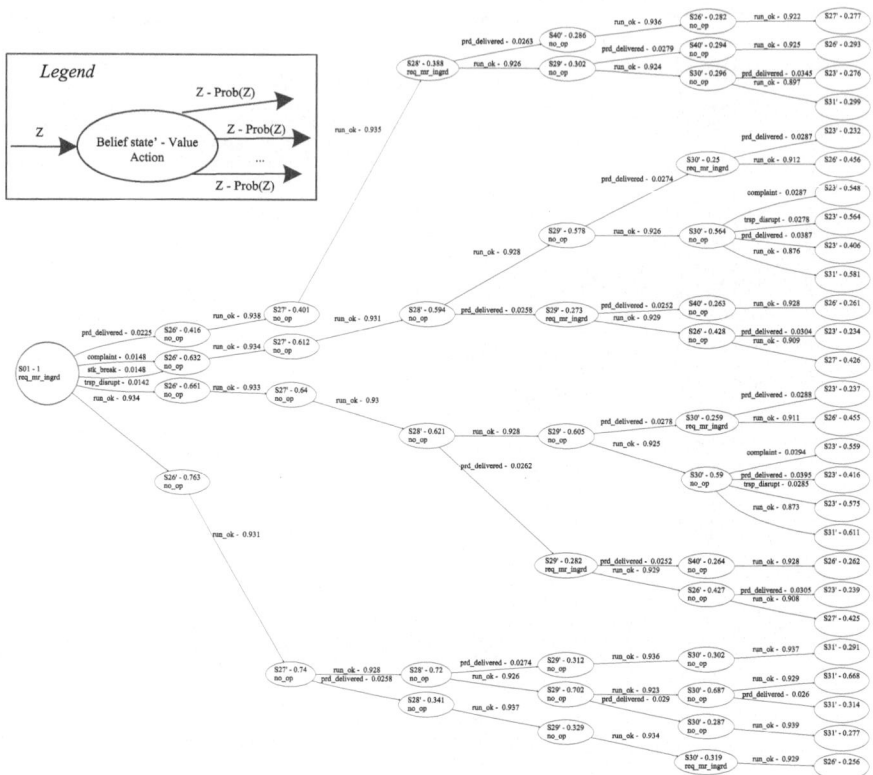

Fig. 6. Agro-food POMDP control policy, *graph-max-depth=6, graph-min-prob=2.5 %*.

engineers are empowered with full pertinent information to forecast the impacts of their decisions in the near future of the organization.

On the one hand, a POMDP delivers a policy graph mapping the observations (Z) into actions (A), maximizing the reward, and yielding the greatest amount of utility over the different decisions through out a time-wide horizon. Whenever a decision is needed, this policy graph guides the engineer. By other words, it serves as a decision map. Unlike the usual challenge of finding the best path in a graph, our solution offers the graph to be followed by the organization. Figure 6 depicts the policy graph with time horizon concerning 6 consecutive observations, whereas the occurring probability is greater than 2, 5 % (except for the first run).

An ellipse represents a belief state (S') that is reached by taking an action (A) and a branch represents an observation (Z). Given the initial state and action the graph follows from the left to the right side, expanding the different actions that are recommended as a reaction for each observation. The value represented in each branch is the probability of occurring a given observation ($Prob(Z)$). The value represented in each ellipse is the actual reward value (**R**) of taking that path throughout execution. Regarding the results delivered by Fig. 6, we identify that the actions that maximize the utility of this specific configuration is given by the path through the actions *no_op*. After 6 consecutive *run_ok* observations

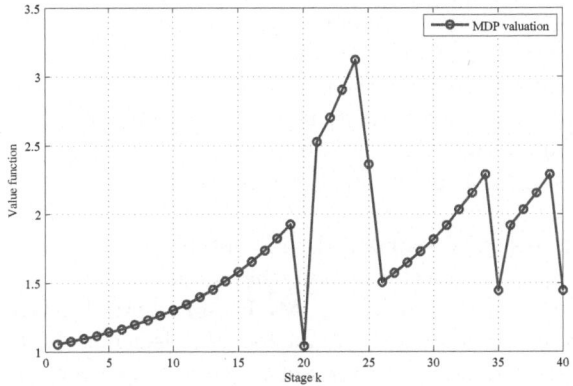

Fig. 7. Agro-food MDP valuation. Each stage is an observable state ($S01...S40$).

the belief state is S31 and the $\mathbf{R} = 0,668$. However, in the case of occurring other observations, Fig. 6 fully describe the rewards for the future actions that should be taken in order to obtain a local maximization.

On the other hand, a MDP solves the problem of calculating an optimal policy in an accessible and stochastic environment with a known transition model. Figure 7 delivers the result of valuating the execution of all consecutive states ($S01..S40$) when each state correspond directly to an observation. S, A, R and γ hold with the definitions contained in Table 2. The transition state T is simpler because only one action is recognized for each state transition. For simulation purposes, the *no_op* action has been considered. In this experimental setup, we find that if no action is taken and the business transactions follows as prescribed the value rise along with the execution of the flows of work.

Comparing POMDP and MDP solutions in terms of benefits for the engineering of decision-making, we find that two different purposes are fulfilled. First, the POMDP results are mapped in a time-wide horizon that forecast the probabilistic belief states from the observations and enacted actions. This result allows the business manager to focus in a black-box perspective of the organization, supporting the decision-making process with more information, even when the business processes are not fully observable. In addition, the business managers are able to dynamically regenerate their strategic plans, whenever any estimation variable or organizational dynamics change. Second, a MDP forecasts local (and global) valuation for business transactions execution assuming that business processes are fully observable. Applying MDP to different business processes design decisions, the optimal (and sub-optimal) solutions that meet the organizational goals could be anticipated prior to its implementation.

6   Conclusions and Future Work

The aim of this paper is to contribute to solve the problem of organizational decision-making (*e.g.*, business processes re-engineering) in partially known environments (usually named by partially observable). Specifically, this paper addresses the environments of business processes execution that are supported by enterprise information systems, which by its turn, are complex and partially observable. The solution support the management decisions, providing maps that express the impacts of management decisions on the organizational operation. Therefore, it minimizes the risk of making wrong decisions (*e.g.*, incorrect change of business processes) and power up a positive impact on the national economy services industry.

To obtain this result, we analyzed the contemporary problems for decision-making and designed a novel solution that combines DEMO-based business process design and operation research (MDP and POMDP Markov theories). In the daily operation, manager and engineers take decisions that are based upon the available observations at each instant in time. Because of partially available information, these observations do not fully describe the actual state of the organization and impose to the manager the problem of guessing what state it is in.

Our solution valuates the actions that could be enacted from the available partial observations, using a probabilistic approach, where an initial estimation effort for the tuple $(S, A, Z, T, O, R, \gamma)$ is demanded. In the end, managers and engineers are empowered with full pertinent information to forecast the impacts of their decisions in the near future of the organization.

Future work will involve two main threads of work: *(i)* the technical integration between the Markov theories and DEMO theory and methodology. On the one hand, the results delivered by this work will benefit from the theoretical POMDP and MDP advances regarding the algorithmic performance optimization and from all the aspects related with fast-approach solution convergence. On the other hand, the estimation of T and O matrices is actually a complex task that demands the development of automatic tools; *(ii)* more case studies are needed in order to achieve a broader generalization of the results and more empirical findings.

Acknowledgments. This work was supported by Project nr 652643 (Respon-SEAble), under the title: *"Sustainable oceans: our collective responsibility, our common interest. Building on real-life knowledge systems for developing interactive and mutual learning media"* from H2020 programme.

References

1. Alter, S.: Theory of workarounds, communications of the association for information systems. Univ. Access Inf. Soc. **34**(55), 1041–1066 (2014)
2. Cai, P.: Approximate pomdp planning (appl) toolkit (2010). https://github.com/petercaiyoyo/appl
3. Cassandra, A.R.: A survey of POMDP applications. In: Working Notes of AAAI 1998 Fall Symposium on Planning with Partially Observable Markov Decision Processes, pp. 17–24 (1998)
4. Cassandra, A.R.: Pomdp solver, v53 (2005). http://www.pomdp.org/code/index.shtml
5. Dietz, J.L.G.: Enterprise Ontology: Theory and Methodology. Springer, Heidelberg (2006)

6. Dietz, J.L.: The deep structure of business processes. Commun. ACM **49**(5), 58–64 (2006). http://doi.acm.org/10.1145/1125944.1125976
7. Dietz, J.L., Hoogervorst, J.A., Albani, A., Aveiro, D., Babkin, E., Barjis, J., Caetano, A., Huysmans, P., Iijima, J., van Kervel, S., et al.: The discipline of enterprise engineering. Int. J. Organ. Des. Eng. **3**(1), 86–114 (2013)
8. Fleischmann, A., Stary, C.: Whom to talk to? a stakeholder perspective on business process development. Univ. Access Inf. Soc. **11**(2), 125–150 (2012). doi:10.1007/s10209-011-0236-x
9. Graphviz: Graph visualization software (2014). http://www.graphviz.org/
10. Group, O.M.: Bpmn specification 1.2. pdf available in internet, January 2009
11. Guerreiro, S.: Decision-making in partially observable environments. In: 2014 IEEE 16th Conference on Business Informatics (CBI), vol. 1, pp. 159–166, July 2014
12. Guerreiro, S.: Business rules elicitation combining markov decision process with demo business transaction space. In: 2013 IEEE 15th Conference on Business Informatics (CBI), pp. 13–20 (2013)
13. Guerreiro, S., Tribolet, J.: Conceptualizing enterprise dynamic systems control for run-time business transactions. In: ECIS, p. paper 5 (2013)
14. Hevner, A.R., March, S.T., Park, J., Ram, S.: Design science in information systems research. MIS Q. **28**(1), 75–105 (2004). http://dl.acm.org/citation.cfm?id=2017212.2017217
15. Hoogervorst, J.A.: Enterprise Governance and Enterprise Engineering. The Enterprise Engineering Series, Springer Science (2009)
16. Laudon, K., Laudon, J.: MIS - Management Information Systems: Managing the digital firm. Pearson, Upper Saddle River (2013)
17. Montiel, L., Bickel, J.: A simulation-based approach to decision making with partial information. Decis. Anal. **9**(4), 329–347 (2012)
18. Puterman, M.L.: Markov Decision Processes: Discrete Stochastic Dynamic Programming. Wiley, New York (1994)
19. Russell, S., Norvig, P.: Artificial Intelligence: A Modern approach, 3rd edn. Pearson, London (2009). ISBN-13: 978-0136042594
20. Shewhart, W.: Economic Control of Quality of Manufactured Product/50th Anniversary Commemorative Issue. American Society for Quality, June 1980
21. Smith, T.: Zmdp software for pomdp and mdp planning (2007). https://github.com/trey0/zmdp
22. Spaan, M.: Perseus randomized point-based approximate value iteration algorithm (2004). http://users.isr.ist.utl.pt/mtjspaan/software/index_en.html
23. Vanderfeesten, I., Reijers, H.A., van der Aalst, W.M.P.: Product-based workflow support. Inf. Syst. **36**(2), 517–535 (2011). doi:10.1016/j.is.2010.09.008
24. Weber, M.: Decision making with incomplete information. Eur. J. Oper. Res. **28**, 44–57 (1987)

Towards Competence-Based Enterprise Restructuring Using Ontologies

Alexey Sergeev[1,2(✉)] and Eduard Babkin[1]

[1] Department of Information Systems and Technologies, National Research University – Higher School of Economics, Bol. Pecherskaya 25, 603155 Nizhny Novgorod, Russia
eababkin@hse.ru
[2] Department of Engineering and Management, Instituto Superior Técnico, Technical University of Lisbon, Lisbon, Portugal
aisergeev@yahoo.com

Abstract. While knowledge of competence is of particular value to the enterprise, and notion of competence is being actively used in enterprise engineering field of research, there is a lack of practical usage of competences while restructuring enterprises. This paper presents a modelling framework for competence-based enterprise restructuring using ontologies, which is based on DEMO's ATD and helps to combine competence requirements on ontological level, competence requirements on implementation level and existing individual competences on implementation level. The number of open questions for future research is being provided.

Keywords: DEMO · Enterprise engineering · Enterprise restructuring · Competence

1 Introduction

It is already few decades that enterprise human resources management shifted from measuring individual productivity toward strategic management of the human resources, with a focus on competence development [1]. Competence management (CM) includes the planning, implementation, and evaluation of initiatives to have sufficient competences of the employees within the enterprise and competence of the enterprise as a whole to reach its goals [2].

Yu et al. [3] state that knowledge about competence is of particular value to the enterprises in such aspects like: understanding is helpful to make correct investment and strategy decisions; increasing awareness of the current industrial norm of capabilities, which can be used to identify areas for further development of enterprise competence; deploying competence to create new markets and new businesses; growing memory of an enterprise; validating the enterprise's ability to undertake some tasks inside or outside the enterprise; publishing the competence of enterprise in the market for potential cooperation opportunity.

Hachicha et al. [4] outline that supervisors or managers have to be aware of the interdependency between competences and tasks to be carried out and the need to

© Springer International Publishing Switzerland 2015
S. Nurcan and E. Pimenidis (Eds.): CAiSE Forum 2014, LNBIP 204, pp. 34–46, 2015.
DOI: 10.1007/978-3-319-19297-0_3

categorize the competences into different resources since it allows a workforce flexible management. It requires an efficient identification of the available competences in the company, a good method to depict the competence needs, an unfailing training strategy. The internal resources and competences of enterprises are increasingly seen as the important factor for dealing with current complex and dynamic environment and acquiring competitive advantages [5]. It is also important to take competences into account while reengineering business processes or doing enterprise restructuring, since right set of competences help to better adapt to concrete environmental conditions. It appeared that existing approaches for enterprise engineering and reengineering do not take competence into consideration, so the large field for research exists.

This paper aims at developing a new framework for competence-based enterprise restructuring using ontologies. Such framework allows to take competences into consideration during enterprise engineering and reengineering, which helps to ensure that no competence is left behind and no competence need left unsatisfied.

This work explored the current research being conducted regarding competences and applied the notion of competence to the field of Enterprise Engineering with a focus on practical usage of competence management. Until now the term competence was well known in the area of enterprise engineering. For example, Dietz [6] considers competence as one of the main attributes of actor, along with authority and responsibility, and defines competence as the ability of a subject to perform particular P-acts as well as the corresponding C-acts. It is of particular importance when applied to enterprise governance [7]. However, there are still very few works related to the practical usage of competence and competence management in enterprise engineering and during its reengineering. The main contribution of this work is in the expansion of traditional ontological-based techniques by adding competences and competence requirements.

The paper is organized as follows: Sect. 2 includes the definition of competence, Sect. 3 provides review of using notion of competence in Enterprise Engineering, Sect. 4 describes proposed framework for competence-based enterprise restructuring using ontologies, Sect. 5 provides an illustrative example of the proposed framework usage, Sect. 6 formulates questions for future research, and Sect. 7 concludes the paper.

2   Definition of Competence

Competence is a term that is used in both everyday language and scientific language. It has a great variety of meanings, but it is possible to extract a core semantic meaning, which corresponds to the terms "ability", "capability", "skill", "effectiveness". This term may have a neutral connotation (it is used to refer to the entire range of an ability from poor to good), or sometimes it may refer only to those abilities that allow good to excellent performance. The term "competence" can be applied to individuals, groups of individuals or organizations. The origin of the word "competence" comes from the Latin, competere. Its meaning is revealed by breaking the word down into com, which means "together" and petere, meaning "effort" [8]; historically it is understood as "cognizance" or "responsibility". In this research authors use the following definition for the term "competence".

Competence is a specialized system of individual or collective abilities or skills that are necessary to perform a particular action.

There is an ambiguous usage of terms "competence" and "competency". Both are sometimes used interchangeably, but there is some difference. Competence is a skill and the attained standard of performance, while competency is the behavior by which a competence is achieved [9]. Competence is a competency in a particular context [10].

It is not necessary for each individual in organization to possess all the competences necessary for successful operation; it is enough when there is a network of competences that allows optimal use of available organizational resources for achieving the goals of the organization. That is why we distinguish three types of competence for closer outlook: individual competence (the set of competences possessed by an individual), group competence (the set of competences collectively possessed by the group of individuals working together), and enterprise competence (entire set of competences collectively possessed by all the employees of an enterprise). It is important to understand that group competence and enterprise competence are not just the set of competences possessed by separate individuals, there may be an effect of synergy when individuals working together on the same task.

3   Notion of Competence in Enterprise Engineering

The term competence was well known in the area of enterprise engineering. For example, Dietz [6] considers competence as one of the main attributes of an actor, along with authority and responsibility, and defines competence as the ability of a subject to perform particular P-acts as well as the corresponding C-acts. However, in corresponding DEMO methodology there is no any practical usage of specific units of competence (skills, knowledge) while building the ontological model of the enterprise and using it for enterprise engineering/reengineering. No any model, while being part of DEMO, includes competence in detail enough for its practical usage.

It is of particular importance when applied to enterprise governance. Henriques et al. [7] use the same definition of competence. In reference method they propose, Step 3 is to "identify the competence domains and define a set of competence principles for each actor role". Competence domains can be perceived as attributes that will guide the evaluation process to check if a person has the adequate competence to exercise its job. Competence principle purposes to restrict the detailed design freedom regarding the actors' production acts. This is the step towards considering competences in the ontological model of the enterprise and using competences while enterprise engineering and reengineering. In the next section development of this idea in a way of framework proposal is provided.

Hoogervorst in [11] provides the definition of competence as 'a coherent whole of organizational skills, knowledge and technology – anchored in the competencies of employees'. Author states that central to the notion of competence is the integration of various enterprise resources. In view of the above, Hoogervorst defines an enterprise competence as an integrated whole of enterprise skills, knowledge and technology. Understandably, competencies must be organized: they are thus an organizational capacity or ability to produce something. And as previously mentioned, integration does not occur spontaneously: intentional activities are required for integration to happen. Author concludes that competence-based view on the enterprise is important and is a base for competence-based governance approach. However, no practical framework for using competences in enterprise engineering and reengineering activities is being proposed.

In our review we found very little to no mentions of practical usage of the notion of competence in relation to the ontology of the enterprise and enterprise engineering. Therefore we believe that this field should be explored in more detail. In the next section we propose framework for competence-based enterprise engineering using ontologies.

4   Proposed Framework for Competence-Based Enterprise Restructuring Using Ontologies

For design of our framework we take basic principles of DEMO methodology and introduce several new concepts. In our research we assume that there are actors in the enterprise which fulfil certain actor roles in order to execute certain (ontological) transactions. Actors possess individual competences and actors are mapped to actor roles contributing their competence to actor role competence on one side. On the other side, there are competence requirements for ontological transactions execution. There may be an inconsistency in requirements and existing competences.

We can draw a map of actors assigned to certain actor roles. Assuming we know individual competences of actors, we can conclude about competences present in actor roles in current implementation of the enterprise.

We can extend this map by adding competence requirements to the actor roles. Competence requirements are dependent on the ontological transactions being initiated and executed by this actor role (assume that initiation of transaction also requires certain

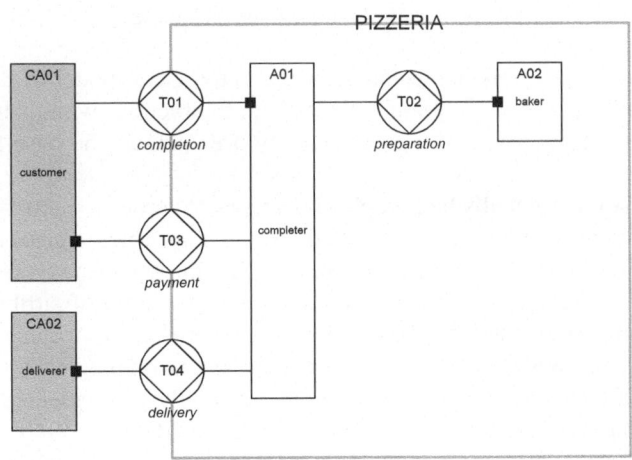

Fig. 1. Original ATD of the Pizzeria, 2nd phase (adopted from [6]).

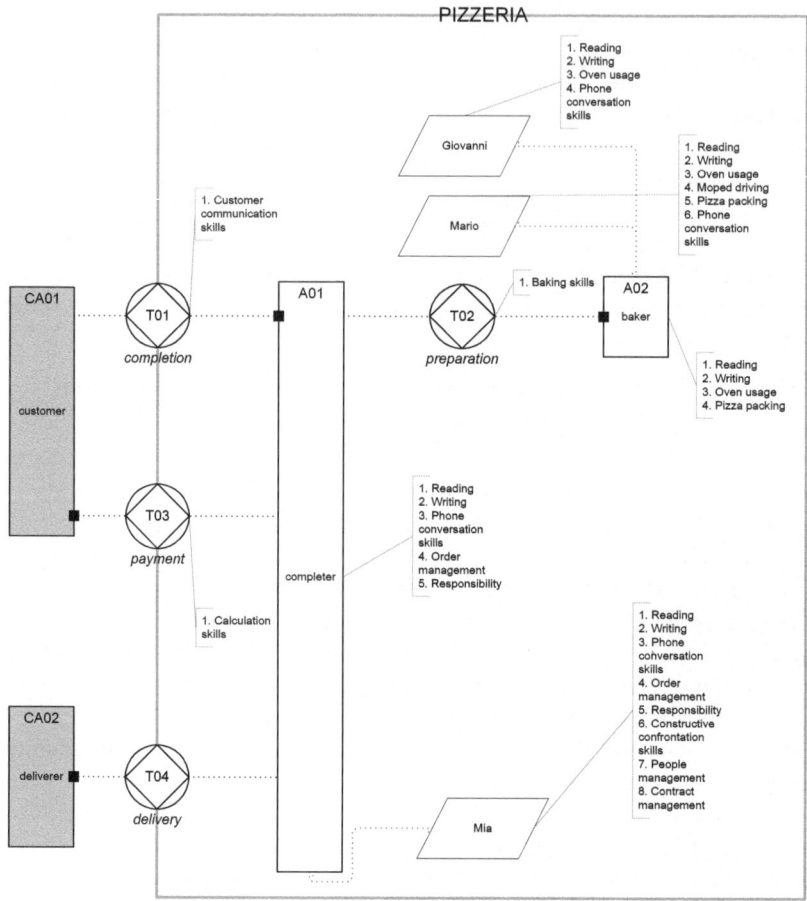

Fig. 2. OCM of the Pizzeria, 2nd phase.

competences) and on the implementation of the ontological model of the enterprise. DEMO's Actor Transaction Diagram includes both transactions with their competence requirements and actor roles, so we can actually use ATD as the base for such map drawing.

Resulting map will actually be a graph with 3 types of vertices – namely actors, actor roles and transactions. Actors are connected only to actor roles, transactions are also connected only to actor roles. There is no direct connection between vertices of the same type and between actors and transactions. Each vertex has a set of attributes – atomic competence items. Attributes for actors are atomic individual competences. Attributes for transactions are atomic competence requirements on the ontological level. Attributes for actor roles are atomic competence requirements on the implementation level. We will call such map an Ontological Competence Map (OCM).

Example of such OCM for the 2nd phase of Pizzeria case (adopted from [6]) – see Fig. 1 for original ATD, Fig. 2 for OCM.

With this map, the framework for competence-based enterprise restructuring is being proposed. Similar to the method described in [12] for considering costs during enterprise restructuring, this framework focuses on competence-based models. The framework consists of ten steps, which leverage usage of the aforementioned competence map and enables enterprise restructuring using ontologies:

Step 1.   Create ontological model of the current state of enterprise using DEMO. Refer to [6] for more details on the DEMO models creation.

Step 2.   Based on the Actor Transaction Diagram, create Ontological Competence Map of the current state of the enterprise. This step requires participation of subject matter experts in order to gather competence requirements for each transaction, participation of business managers to gather information about current assignment of employees to actor roles, and information about individual competences of employees (this information can be gathered from competence management systems, if they are implemented in the enterprise).

Step 3.   Create ontological model of the planned state of the enterprise after restructuring using DEMO.

Step 4.   Based on the Actor Transaction Diagram, create Ontological Competence Map of the planned state of the enterprise after restructuring. At this step, OCM will be complete in terms of ontological competence requirements, but may be incomplete for implementation competence requirements and individual employee competences.

Step 5.   Compare OCMs produced on Step 2 and 4 in order to understand:

- which new ontological requirements are added to the model and not yet fulfilled;
- which individual competences are removed from the model because of employees leaving the company or changing their roles;
- what are the changes for implementation competence requirements.

Step 6.   Analyze OCM produced on Step 4 with the information obtained on Step 5 in order to finalize implementation competence requirements.

Step 7.   Reassign employees between actor roles based on their individual competences in order to fulfil implementation competence requirements in the most optimal way, where possible. The main aim of this step is to optimize employee functions based on their individual competences, and to minimize number of new employees to be hired. Produce new OCM.

Step 8.   Based on the OCM produced on Step 7, finalize number of people to be hired and competence requirements for them. Produce final OCM for the planned restructuring.

Step 9.   Hire new employees based on the requirements produced on Step 8 and assign them to the actor roles as per the final OCM.

Step 10.   Produce OCM of the state of the enterprise after restructuring ends to validate that no competence gaps exist. If competence gaps exist, proceed with additional employee hiring or employees' trainings to remove the gaps.

Practical relevance of OCM for current enterprise operation is to help understand the lack of competence in order to do reassignment of employees to actor roles, plan employee trainings or hire new employees (and, possibly, fire some existing employees). OCM also helps to optimize employee assignment to actor roles to minimize number of employees to be hired. In case new employees are needed to be hired, OCM shows all competence requirements for new hires.

Proposed framework is relevant for enterprise restructuring, because it ensures that no gaps in competence will exist in new structure, helps to plan budgets for employee hiring and training, and helps to build optimal implementation of the ontological model of the enterprise in terms of enterprise competence.

Next section includes illustrative example of the framework usage for aforementioned Pizzeria case adopted from [6].

5   Example of Framework Usage

This section shows the illustrative usage of the proposed framework for competence-based enterprise restructuring using ontologies for the Pizzeria case, adopted from [6]. Assume that current state of the enterprise is Phase 2, and restructuring is planned to achieve Phase 3.

Step 1.   See [6] for full ontological model of the Pizzeria at the Phase 2. Figure 1 shows the ATD of the Pizzeria, which is used on next step to produce OCM.
Step 2.   See Fig. 2 for OCM of the Pizzeria at the Phase 2.
Step 3.   See [6] for full ontological model of the Pizzeria at the Phase 3. Figure 3 shows the ATD of the Pizzeria, which is used on next step to produce OCM.
Step 4.   See Fig. 4 for OCM of the planned Pizzeria structure at the Phase 3. For the sake of simplicity, let's assume that it was already possible to plan implementation competence requirements and include into this OCM. Note that both Mia and Mario are leaving the company, so no any employees are actually shown fulfilling any actor roles.

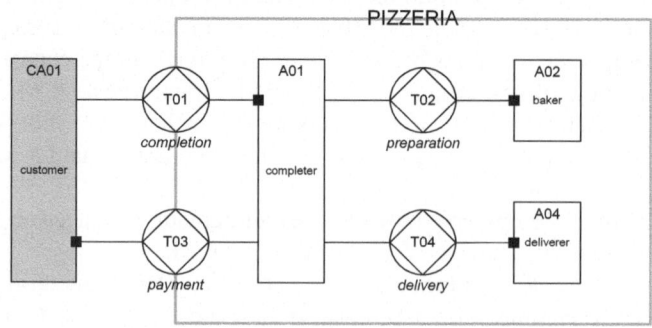

Fig. 3.   Original ATD of the Pizzeria, 3rd phase (adopted from [6]).

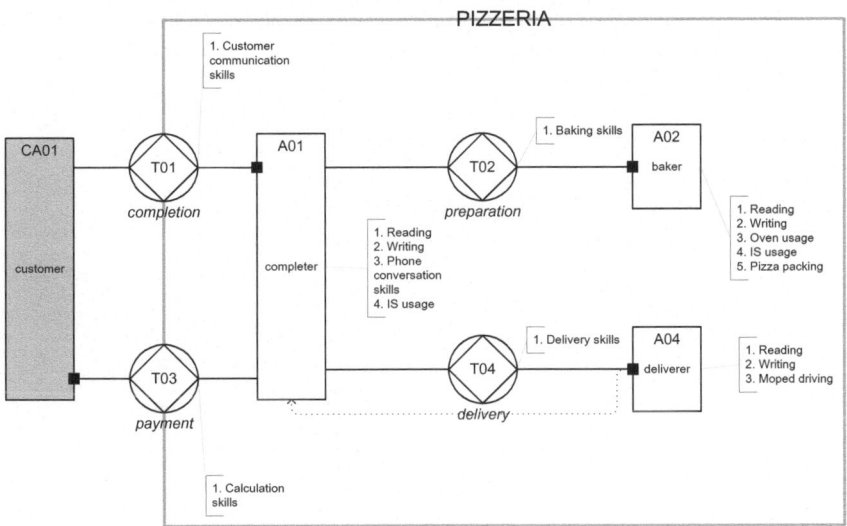

Fig. 4. OCM of the planned Pizzeria structure at the Phase 3.

Step 5 and 6. Comparing Figs. 2 and 4, it can be seen that no any employees are currently fulfilling any ontological actor roles. Also, "Information Usage skills" were added to the implementation competence requirements for Completer and Baker actor roles. To add to this, as new transaction – T04. Delivery – was added to the ontological model, and there is new ontological competence requirement – "Delivery skills". To fulfill this requirement, three new implementation competence requirements were added for Deliverer actor role.

Step 7. In this particular case, as no any employees available for reassignment (assume that Giovanni is fully occupied with his new position), this step is skipped.

Step 8. As an example, we consider that planned implementation on the Phase 3 requires two bakers and 2 deliverers to be hired. Final OCM for the planned restructuring to Phase 3 is shown on Fig. 5. Note that this OCM helps to represent the hiring requirements, and for each employee to be hired the competence requirements are known. E.g. for the deliverer they are: customer communication skills, calculation skills, reading, writing, and moped driving skills.

Step 9. Assume that owners hired two deliverers and two bakers with best match to implementation competence requirements. For the illustration purposes, let's call these new employees as "deliverer 1", "deliverer 2", "baker 1" and "baker 2". We assume that some new hires do not fully satisfy competence requirements (which is typical for the labor market).

Step 10. OCM of the Pizzeria at the phase 3 is shown on Fig. 6. Analyzing this OCM, we can conclude that Deliverer 1 lacks IS usage skills, Deliverer 2 lacks phone conversation skills, Baker 1 fully satisfies competence requirements and has additional skill of moped driving, which may be useful in case any further restructuring planned, Baker 2 lacks pizza packing skills. So, the recommendation to the management of the Pizzeria at the Phase 3 will be to train employees to remove competence gaps. Desired OCM of the Pizzeria should then look like shown on the Fig. 7.

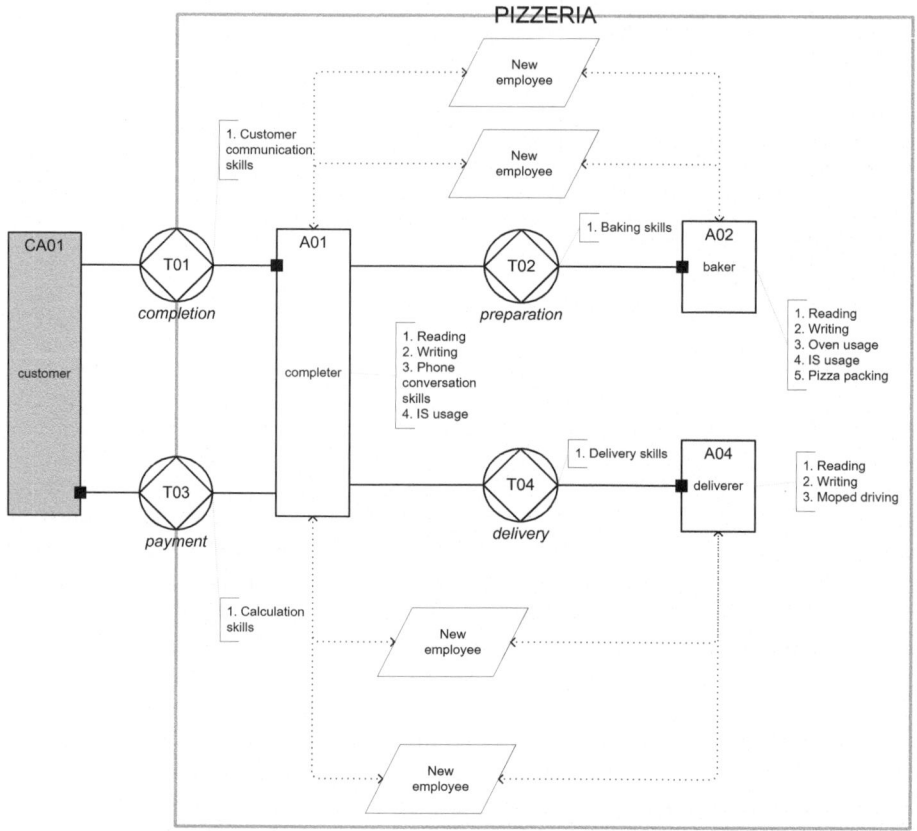

Fig. 5. Final OCM of the planned Pizzeria structure at the Phase 3.

6   Future Research

In order to effectively use competences while reengineering business processes or restructuring an enterprise, a number of questions need to be answered. This section includes questions for future research which authors of this paper plan to explore.

The work with competences at the enterprise starts with the individual competences, and the first question arises – how to retrieve the set of competences which an individual possesses? It is important to get the exhaustive list of individual competences, and it is likewise important that this list must not be excessive. We rely on existing techniques from psychology and decision theory to answer this question. Considering that the

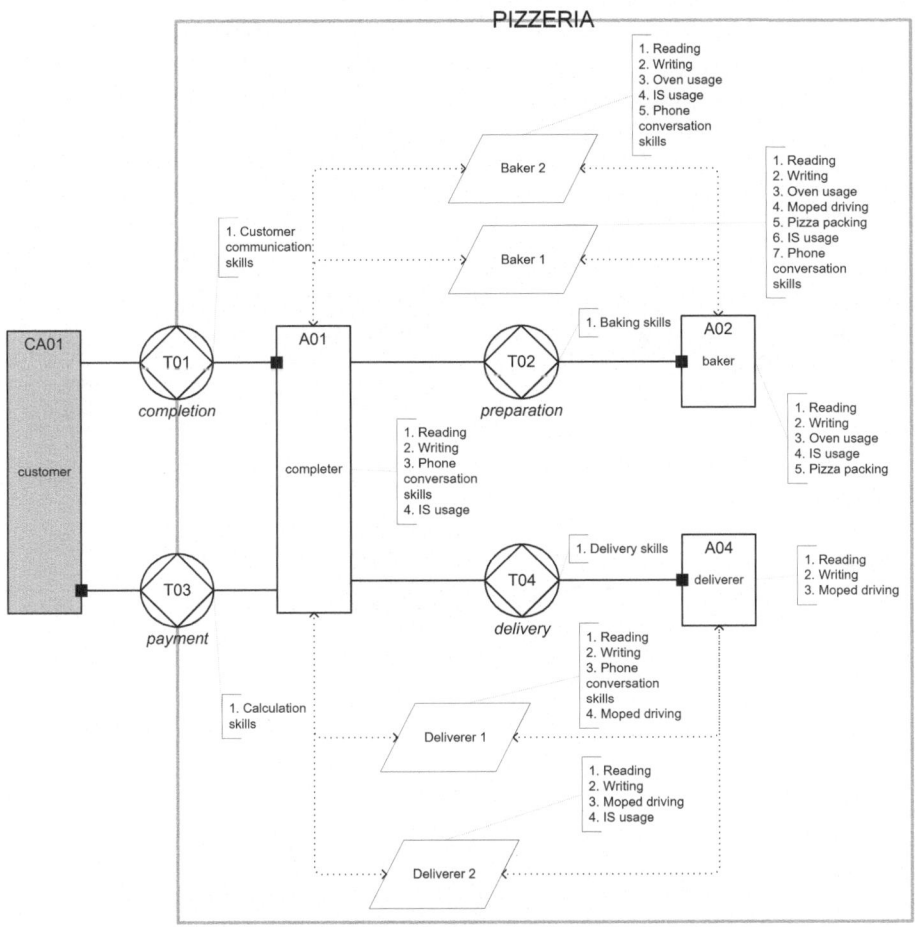

Fig. 6. OCM of the Pizzeria, 3rd phase.

exhaustive list of individual competences was retrieved, the next question is how to model individual competences in a way appropriate for further analysis? Obviously, having the list of competences does not mean that one is able to analyze it and get any useful information out of it. This set of individual competences needs to be codified and represented in a way appropriate for analysis. At this stage, we need to separate two possible types of analysis to be done with individual competences – manual analysis and automated analysis. Manual analysis is to be done by human, therefore, we need such a method to model individual competences so that human is able to correctly understand them and operate them to do analysis. Automated analysis means analysis done by machine, therefore, we need such a method so that set of competences can be

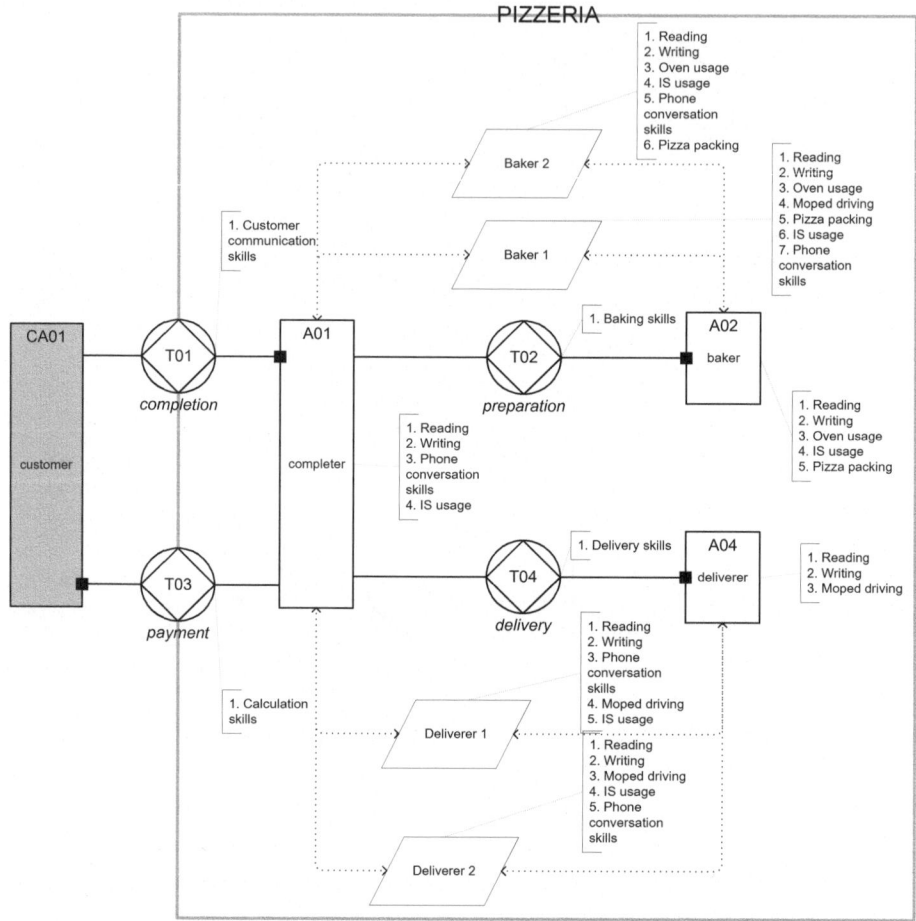

Fig. 7. Desired OCM of the Pizzeria after employee trainings completed.

entered to the computer software, stored by this software, interpreted by software, and "analyzed" by software.

Framework for competence-based enterprise restructuring using ontologies, presented in Sect. 4, solves the problem of individual competences representation for manual analysis. However, it does not answer the outlined question of automated analysis. To add to this, once modelling methods for both manual (provided in Sect. 4) and automated types of analysis are created, the methods of analyses per se should be proposed.

Finally, once modelling and analysis methods for individual, group and enterprise competences exist, software for automation of modelling and analysis of competences can be created. While one can find many competence management software systems, none of them are working with ontological model of enterprise. Therefore, one of the questions for future research is the creation of specification for such a competence management software, which would make possible to automate all processes mentioned in research questions above.

7 Conclusion

In this paper we introduce the definition of competence which later be used in further research. The notion of competence in the field of enterprise engineering was studied. While competence is being used in the theory of enterprise engineering and enterprise ontology, the lack of practical usage of competences during enterprise engineering and reengineering was identified.

As a result of this, the modelling framework for competence-based enterprise restructuring was proposed. It maps ontological model of the enterprise to its implementation in terms of competence requirements and existing individual competences. The framework is built upon DEMO's Actor Transaction Diagram extended with actors being added, as well as existing competences and competence requirements on both ontological level and implementation level. This framework addresses the problem of competences and competence requirements modelling and representation, which will be used for further analysis.

The number of questions for further research in this field were provided. These questions include retrieving of information about individual competences, modelling of competences (which wasn't yet solved for computer-based modelling), analysis of competences (including both manual and automated analysis).

References

1. Munkvold, B.E., Hustad, E.: IT-supported competence management: a case study at ericsson. Inf. Syst. Manage. **22**(2), 78–88 (2005)
2. Nordhaug, O.: Human Capital in Organizations. Scandinavian University Press, Oslo (1993)
3. Yu, L., Biqing, H., Wenhuang, L., Hongmei, G., Cheng, W.: Enterprise Competence Modeling and Management. 0-7803-6583-6/00/$10.00 0 2000 IEEE
4. Hachicha, R.M., Dafaoui, E.M., Mhamedi, A.E.: Competence evaluation approach based on 2-tuple linguistic representation model. 978-1-4244-3672-9/09/$25.00 ©2009 IEEE
5. Prahald, C.K., Hamel, G.: The core competence of the corporation. Harvard Bus. Rev. **68**(3), 79–90 (1990)
6. Dietz, J.L.G.: Enterprise Ontology: Theory and Methodology. Springer, Heidelberg (2006). ISBN-10 3-540-29169-5
7. Henriques, M., Tribolet, J., Hoogervorst, J.: Enterprise governance and DEMO - Guiding enterprise design and operation by addressing DEMO's competence, authority and responsibility notions. In: Barbosa, L.S., Correia, M.P. (eds) INForum 2010 - II Simposio de Informatica, pp. 473–476, 9–10 Setembro 2010
8. Czelusniak, J., Abreu, A., Dergint, D., Hatakeyama, K.: Proposal of agent's software for support competence management process. In: PICMET 2010 Conference (2010)
9. Sanghi, S.: The Handbook of Competence Mapping: Understanding, Designing and Implementing Competency Models in Organizations. Sage Publications, London (2004)

10. Vervenne, L., Najjar, J., Ostyn, C.: Competency Data Management (CDM), a proposed reference model. In: The European Competency SIG workshop, Berlin, December 2008
11. Hoogervorst, J.A.P.: Enterprise Governance and Enterprise Engineering. The Enterprise Engineering Series, 1st edn. Springer, Heidelberg (2009)
12. Babkin, E., Sergeev, A.: Towards developing a model-based decision support method for enterprise restructuring. In: Proper, H.A., Aveiro, D., Gaaloul, K. (eds.) EEWC 2013. LNBIP, vol. 146, pp. 17–27. Springer, Heidelberg (2013)

The Enterprise Engineering Domain

Marne de Vries[1(✉)], Aurona Gerber[2], and Alta van der Merwe[3]

[1] Department of Industrial and Systems Engineering,
University of Pretoria, Pretoria, South Africa
marne.devries@up.ac.za
[2] Centre for Artificial Intelligence Research, CSIR Meraka,
Department of Informatics, University of Pretoria, Pretoria, South Africa
agerber@csir.co.za
[3] Department of Informatics, University of Pretoria, Pretoria, South Africa
alta.vdm@up.ac.za

Abstract. Enterprise engineering (EE) is emerging as a new discipline to address the design of the enterprise in a holistic way. Although existing knowledge on enterprise design is dispersed and fragmented across different disciplines and approaches, previous research presented an enterprise evolution contextualisation model (EECM) as a representation of the existing EE body of knowledge. Since EECM was developed inductively from existing design/ alignment/governance approaches, EECM was also proposed as a representation of the EE domain within the emerging EE discipline. We used a questionnaire to gather the views of EE and enterprise architecture (EA) researchers and practitioners on the EE domain. The main contributions of this article include: (1) the validation results of the proposed boundaries of the EE domain, and (2) a prioritisation of the phenomena of interest and core problems or topics of interest within the EE domain.

Keywords: Enterprise engineering · Enterprise engineering discipline · Enterprise engineering research agenda

1 Introduction

An enterprise originates when man and machine are organised to pursue some common goal [1]. Researchers and practitioners from different disciplines study the enterprise as a phenomenon, but also contribute to its evolution. Disciplines such as systems engineering, industrial engineering, information systems, management sciences, psychology, sociology and organisational sciences all contribute from different perspectives to understand, design and engineer the enterprise [1, 2].

Both researchers and practitioners expressed the need for a comprehensive view of the enterprise in different publications [1, 3–5], culminating in the emergence of a *new discipline*. The new discipline, *enterprise engineering* (EE), could be defined as "the body of knowledge, principles, and practices to design an enterprise" [1], consisting of three subfields, namely enterprise ontology, enterprise governance and enterprise architecture [6].

© Springer International Publishing Switzerland 2015
D. Aveiro et al. (Eds.): EEWC 2015, LNBIP 211, pp. 47–63, 2015.
DOI: 10.1007/978-3-319-19297-0_4

EE as a young discipline is often regarded as an extension of the fields of industrial engineering or business process management, consequently "the current status of enterprise engineering initiatives, as taken by several universities, is unclear" [7, p. 93]. Furthermore, a plethora of EE-related literature from different disciplines exist, but with a lack of shared meaning [8]. The lack EE description and terminology led to a research initiative to describe the existing body of knowledge via a common reference model, the enterprise evolution contextualisation model (EECM) [9]. EECM was inductively developed from existing enterprise design/alignment/governance approaches and EECM therefore presents a high-level meta-model for the existing body of knowledge within EE [10].

A key prerequisite for establishing a new discipline is to *define the domain* of the discipline answering three fundamental questions: (1) what phenomena are of interest in the discipline? (2) what are the core problems or topics of interest? and (3) what are the boundaries of the discipline [11]? This article has two main objectives, firstly to *validate* if EECM represents the scope of the EE domain and secondly to *prioritise* the phenomena of interest and core problems or topics of interest within EE.

The article is structured as follows: Sect. 2 provides background theory on EECM and the proposed *domain of EE* related to the four EECM components. Section 3 presents the research method, i.e. using a questionnaire to gather opinions on the pre-defined *domain of EE*, followed by the questionnaire results in Sect. 4. In Sect. 5 we discuss the research results and provide suggestions for future research in Sect. 6.

2 Background Theory

Previous research highlighted the fragmentation that exists within the EE discipline and the need to provide a common reference model to understand and compare existing knowledge within the EE discipline [12]. EECM was developed inductively from existing enterprise design/alignment/governance approaches, as a common reference model to contextualise/translate an existing approach [9, 12, 13]. Section 2.1 presents EECM, followed with the suggestion in Sect. 2.2 to use EECM as a means for demarcating the domain of the EE discipline.

2.1 The Enterprise Evolution Contextualisation Model (EECM)

When any researcher or practitioner uses EECM (depicted in Fig. 1) to contextualise a specific enterprise design/alignment/governance approach, EECM asks three main questions about the specific approach:

- Question 1: '*Why* should the enterprise use the proposed approach to evolve?'
- Question 2: '*What* should the enterprise evolve?'
- Question 3: '*How* should the enterprise evolve?'

In answering the three questions through a conceptual mechanism, EECM subsequently consists of four main components (Fig. 1).

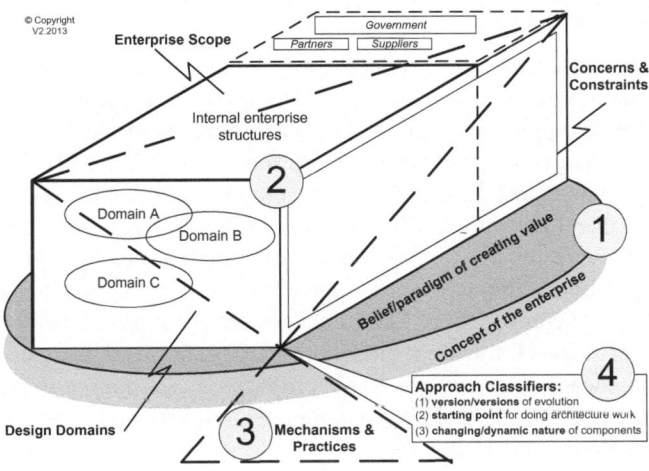

Fig. 1. The enterprise evolution contextualisation model (EECM)

2.2 The Domain of the Enterprise Engineering Discipline

Gregor [11] suggests that we consider three questions to define the domain of a discipline namely (1) What phenomena are of interest in the discipline? (2) What are the core problems or topics of interest? and (3) What are the boundaries of the discipline?

Since EECM was developed inductively (bottom-up) from an extensive analysis of current prevalent EE approaches, EECM represents a high-level categorisation and meta-model of the exiting EE body of knowledge. As proposed in De Vries et al. [10], the four components of EECM could be used to answer the three questions pertaining to the *domain* of the EE discipline. Table 1 summarises the relationship between the domain questions and the four EECM components. A motivation for each relationship is then discussed.

Table 1. Relating the domain questions to the EECM components

Domain questions	EECM components
What phenomena are of interest in the discipline?	Component 1: Concept of the enterprise & paradigm of creating value
What are the core problems or topics of interest?	Component 2: Dimensions
	Component 3: Mechanisms & practices
	Component 4: Approach classifiers
What are the boundaries of the discipline?	Component 2: Dimensions
	Component 3: Mechanisms & practices
	Component 4: Approach classifiers

What Phenomena are of Interest in the Discipline?

Our era is characterised by rapid technological change, including connectivity; smart devices and ubiquitous computing; and the generation of and access to a vast amount of information. The enterprise as a socio-technical system is in the epicentre of the impact of most of these changes. Changes that modern society experience also have an impact on the observed phenomena that characterises the discipline of EE [10].

During the development of EECM, there was evidence that EE researchers addressed phenomena related to the above mentioned discussion. Multiple authors proposed design/alignment/governance approaches to address the enterprise-related phenomena. The first component of EECM acknowledges approach authors' *concept of the enterprise* or way of thinking in defining the enterprise. In addition, the first component also encapsulates authors' value propositions (*paradigm of creating value*) that are presented as part of their design/alignment/governance approach to address/ solve phenomena/problems [10].

Table 2 includes phenomena of interest that are represented in existing EE literature [10, 12]. From a futuristic perspective, we also included phenomena that are currently emerging and may have an impact on the design of the enterprise in future [14]. In Sect. 4.4 we presented additional EE phenomena that were identified by the respondents. If the additional phenomena could be associated with those in Table 2, we included the respondent comments in italics.

What are the core problems or topics of interest?

When observed *phenomena* within any discipline are analysed, distinct domain problems are identified. The content of existing enterprise design/alignment/governance approaches reveal several topics of interest, which have been consolidated in the second, third and fourth components of EECM [10]. Table 3 provides a list of EE problems/topics of interest that were extracted from several approaches [9, 12]. In Sect. 4.5 we presented additional EE problems or topics of interest that were identified by the respondents. If the additional problems or topics of interest could be associated with those in Table 3, we included the respondent comments in italics.

What are the boundaries of the discipline?

The *EE phenomena of interest* and *EE problems/topics of interest* that have been discussed in this section already demarcate the boundaries of the EE discipline. The boundaries are also discussed extensively in [10] and can be summarised as follows:

- Component 1 - (Concept of the enterprise and paradigm of creating value): creates a philosophical boundary.
- Component 2 - (Dimensions): creates a design scope boundary.
- Component 3 - (Mechanisms and practices): creates a practical facet boundary (different "ways of").
- Component 4 - (Approach classifiers): creates an approach pattern boundary, i.e. different approach preferences.

Table 2. Phenomena of interest for EE, based on [10, 14]

No.	Phenomenon description
1	The enterprise as a complex socio-technical system struggles to adapt swiftly to rapidly-changing environments, technologies and customer expectations regarding innovative/variety of products/services. How could enterprises continuously transform from an existing state to a future state in an agile way?
2	Due to enterprise complexity, holistic approaches are lacking to view/understand business processes, information, enterprise applications and technology. How could a holistic view/understanding of the enterprise assist with intelligent investment decisions?
3	Enterprises fail to implement strategic initiatives successfully, which is primarily due to the lack of enterprise governance. How could enterprises create governance (coherence and consistency among the various components of the enterprise)?
4	Enterprises receive limited value from IT investments, which is primarily due to the lack of IT governance. How could enterprises better align IT systems with business requirements?
5	Multiple concerns and interests of enterprise designers (decision-makers) are not-integrated, not-transparent and conflicting, causing ineffective enterprise design. How should multiple enterprise interests or concerns be integrated or aligned? *"Enterprise design should consider the biases of the system, the ever changing aspects which is the people. As long as people are core to the socio-technical system of the enterprise, business problems will never be simplistic."* *"Shareholder value concerns are not aligned with building a sustainable profitable enterprise as it causes short term behaviour and artificial gains that obviate true value generation."*
6	Large data sets may have to be shared/utilised beyond the scope of the enterprise (when the enterprise is considered as a single legal entity), which creates challenges in terms of data ownership. How could enterprise design address this challenge?
7	Increasing heterogeneity of work forces and collaboration between employees and clients across boundaries and across the globe creates new challenges regarding power, roles and responsibilities. How could enterprise design address this challenge?
8	Customers demanding always-availability of enterprise services (made possible due to modern ICT technologies) impose new challenges on the operation of the enterprise. How could enterprise design address this challenge?
9	Society's concern with miss-management of natural resources and ill-treatment of low-income labour requires enterprise design that incorporates design for a triple bottom line (social, environmental and financial). How could enterprise design address these 3 concerns concurrently?

We conclude that EECM could be adopted as a mechanism for answering the questions regarding the *domain* of the EE discipline. The next section presents the research method for *validating* the domain of EE and *prioritising* the phenomena of interest and core problems or topics of interest.

Table 3. Problems/topics of interest for EE based on [10, 14]

No.	Core problems or topics of interest
1	There is a need to define design domains. Different conceptualisations exist for demarcating the enterprise into design domains or components (examples of design domains include: business, organisation, information, and technology) *"Identifying information needs (not IT) as part of the EE effort. The 2 main topics in cooperation between responsible parties are: cooperation by "doing business" (based on negotiable mutual commitments) and information exchange (which also requires commitments, but at a different level)."* *"We must focus more on understanding the assumptions that bias our understanding of enterprises. We must also focus more in learning and transferring knowledge from other disciplines."* *"For which aspects of enterprises is formal design fruitful?"* *"Avoiding theoretical and (as a consequence) practical fragmentation in addressing enterprises."*
2	There is a need to define/align concerns and interests (functional and non-functional) that should be addressed by the entire enterprise and its demarcated design domains. Examples of enterprise functional concerns include production/rendering of products/services to markets. Examples of enterprise non-functional concerns include robustness, agility, flexibility and scalability *"Understanding the stakeholders within the enterprise."*
3	There is a need to define existing constraints that restrict design of the entire enterprise and its demarcated design domains. Examples of constraints include physical limitations of resources or regulatory rules
4	There is a need to define the extent/scope of design/alignment/governance in terms of internal enterprise structures. Examples of existing internal enterprise structures include business units, departments, programs or projects
5	There is a need to define the extent/scope of design/alignment/governance in terms of internal enterprise structures AND extended enterprise structures. Extended enterprise structures include organised structures of partners, suppliers/supply chain, government etc
6	There is a need for architecture description, reference models and modelling practices for different enterprise design domains. Architecture description usually manifests as architecture description languages, such as DEMO or BPMN or as architecture frameworks, such as the Zachman Framework. Generic reference models are used to quick-start architecture efforts, such as e-TOM and SCOR. Modelling practices provide additional guidance when applying an architecture description language
7	There is a need for selection and measurement of concerns and interests. Multiple stakeholders are involved during enterprise design. Mechanisms are required to select appropriate concerns and interests for the entire enterprise and its demarcated design domains. Performance measurement mechanisms are required to measure against the stated concerns and interests. An example of a mechanism to select and measure concerns is the Balanced Scorecard *"The measures of success need to be re-evaluated. Traditional financial measurements often stimulate the wrong behaviour for optimised performance. They often stimulate sub-optimised behaviour."*

(Continued)

Table 3. (*Continued*)

No.	Core problems or topics of interest
8	There is a need for design methodologies for designing the entire enterprise. The scope of a methodology could be defined in terms of the object/entity/sub-system that needs to be designed. Some methodologies focus on the design of a single domain/ sub-system, e.g. the information domain. Yet, there is a need for methodologies that facilitate design of the enterprise as a holistic entity *"Enterprise Engineering needs to focus on "engineering = creating" a better enterprise rather than trying to just be a new term for stringing Enterprise Governance and Enterprise Architecture under one umbrella. Enterprise Engineering also needs to be directed by the Enterprise Architecture and design thereof i.e. EE should follow after EA happened."* *"Is comprehensiveness of enterprise engineering theory and methods necessary and sufficient for designing enterprises in a unified and integrated manner?*
9	There is a need for enterprise design methodologies that cover the entire enterprise life cycle. The scope of a methodology could be defined in terms of the entity's 7 life cycle phases: entity identification, entity requirement/analysis, entity design, entity construction, entity implementation, entity operation & maintenance, and entity decommissioning
10	There is a need for architecture governing principles to ensure unified and integrated design. Architecture principles are general rules and guidelines that support the way in which an enterprise intends to fulfil its mission. The phrase 'decision criteria' is also used as a synonym for shared principles that guide decision-making during enterprise design
11	There is a need for governing mechanisms, practices (practice frameworks) & standards. Governing mechanisms are required within existing management areas to ensure coherent and consistent evolution of the enterprise. Existing management areas include architecture management, change management, strategy management, risk management, program & project management, requirements management, configuration management, quality assurance. Required governance practices are often embedded in frameworks, such as SCOR (Supply-Chain Operations Reference-model). Prescribed standards are needed to constrain design freedom, especially during the design of the ICT system, e.g. the Standards Information Base developed by the Open Group *"Value chain and supply circle thinking combined with project engineering skill could prove useful to add overall context to EE practice."* *"Given the nature of the enterprise, what are adequate governance practices?"* *"Capabilities: what capabilities do an enterprise need and how can these be designed and dynamically adapted. How can capabilities interact effectively?"* *"Embedding EE in the current way-of-working of the enterprise. Getting it appreciated and applied."*
12	There is a need for transformation roadmaps. Roadmaps list individual increments of change according to a timeline to show progression from the current- to future state. Roadmaps are often encapsulated in program management or project portfolio management, where multiple projects are identified for closing the gap between the current state and future state

(Continued)

Table 3. (*Continued*)

No.	Core problems or topics of interest
13	There is a need for problem analyses prior to enterprise re-design. Problem analyses practices and mechanisms facilitate the identification of problems and the severity/effects of existing problems
14	There is a need for gap analyses during enterprise re-design. A gap analysis identifies the gap between the current-state architecture and future-state architecture of an enterprise.
15	There is a need for impact analyses during enterprise re-design. An impact analysis is used to estimate the impact/feasibility of different alternative solutions in terms of cost, schedule or change impact
16	There is a need for enterprise maturity assessment. Different maturity models exist to assess the level of maturity within an enterprise. There is a need to assess IT governance maturity, but also enterprise governance maturity
17	There is a need for specifying skills/learning requirements for developing EE capabilities within an enterprise. Enterprise design/alignment/governance requires skilled employees. Different skills frameworks exist for defining EE skills
18	There is a need for EE software tools to perform consistent enterprise modelling. A wide variety of tools and tool sets are available to perform enterprise modelling. No single tool provides consistent modelling across all enterprise design domains
19	There is a need for tools to guide decision-makers in selecting EE software tools to model the enterprise or a specific design domain. Criteria and guidelines exist for comparing and choosing EE software tools for enterprise modelling
20	There is a need to understand design/alignment/governance approaches in terms of their focus on a specific version of evolution. Some approaches focus on describing and understanding the current (as-is) version of the enterprise. Others focus on the future (to-be) version of the enterprise
21	There is a need to understand design/alignment/governance approaches in terms of their starting point for modelling the enterprise. Approaches focus on different enterprise domains during modelling. Some approaches promote a top-down approach, starting with the business organisation system (top level) prior to modelling the ICT organisation system (bottom level). Others focus on re-designing the bottom level using flexible IT infrastructure to accommodate business (top level) changes
22	There is a need to understand design/alignment/governance approaches in how they address the dynamics of an ever-changing enterprise. Approaches propose different means for addressing the dynamic nature of architecture components

3   Research Method and Data Collection

Since EE is an emerging discipline, the *domain of EE* should be validated by practitioners and researchers that are active within the discipline. We used a questionnaire as a data-gathering instrument to source opinions from EE/EA researchers and practitioners.

EECM provides a new classification scheme for defining the *EE domain* [10] and has not yet been adopted as a representation of the EE body of knowledge. We used several opportunities to present the new classification scheme to EE researchers and practitioners during three different events, firstly the Enterprise Engineering Working Conference in May 2014, secondly the Enterprise Architecture Research Forum meeting in June 2014, and thirdly the Conference of the South African Institute for Industrial Engineering in July 2014. The presentations introduced the *EE domain* (extracted from the EECM components) to the event attendees, where after attendees and other EE researchers and/or practitioners were invited to complete an electronic questionnaire.

The *dependent variable* is the *domain of EE*. The independent variables are factors that influence the dependent variable, i.e. the opinions regarding the domain of EE. Since respondents differ in training and experience within enterprise engineering, the *respondent context* also influence opinions regarding the *domain of EE*. This research is not aimed at quantifying the cause-and-effect relationship between the dependent and independent variables, but acknowledges *respondent context* when interpreting the questionnaire results. Since the target group of participants are practitioners within EE, the *respondent context* was also used to remove results obtained from participants that fell outside the target group.

Three sections gathered data pertaining to the *respondent context*:

- Academic background: Gathering data about the respondent's highest qualification and the tertiary education institution.
- Skill focus: Gathering data about the respondent's EE and EA skills, based on the EA knowledge areas defined by Hoogervorst [5] and the architecture skills framework of TOGAF 9.1 [15]. However, we excluded the detailed set of technical IT skills presented by TOGAF 9.1, since the intention was not to analyse the cause-and-effect relationships between the *respondent context* and the opinions about the *domain of EE*. We have also excluded the generic skills and traits since these do not highlight the focus areas within EE/EA.
- Industry background: Gathering data about the working environment (type of industries) and different EE/EA roles taken in industry.

Three sections gathered opinion data regarding the *domain of EE*:

- Phenomena of interest.
- Core problems or topics of interest.
- The boundaries of the EE discipline.

For the opinion data, we requested that respondents rated the necessity/importance of pre-identified *phenomena of interest, core problems or topics of interest*, and ways of specifying the *boundaries of the discipline*. The questionnaire also included open-ended questions to incorporate additional phenomena of interest and problems or topics of interest. Respondents also had to comment on the ability/inability of using the pre-defined EE boundaries as a means to demarcate the boundaries of the EE discipline.

4   Results

This section reports on the questionnaire results regarding the *respondent context* in Sects. 4.1, 4.2 and 4.3 and the opinions about the *domain of EE* in Sects. 4.4, 4.5 and 4.6.

4.1   Academic Background

A total number of 22 participants completed the questionnaire during the period 12 September to 1 October 2014. In terms of academic background, 15 out of 22 respondents (68 %) obtained a research-related post-graduate qualification, i.e. 36 % obtained a Masters, whereas 32 % obtained a PhD.

Respondents received qualifications from 13 different international universities: University of Pretoria (South Africa), University of Stellenbosch (South Africa), University of South Africa, University of Johannesburg (South Africa), University of Science and Technology (Zimbabwe), University of Montreal (Canada), University of Toronto (Canada), University of Twente (Netherlands), Endhoven University of Technology (Netherlands), Radboud University (Netherlands), University of Amsterdam (Netherlands), Technical University Delft (Netherlands), and Utrecht University (Netherlands). The majority of respondents completed their tertiary education at South African universities.

Respondents also had to indicate the topic of their dissertation/thesis. Based on the descriptions provided by the respondents, the dissertation/thesis topics were associated with the 22 *topics of interest* that were specified in the survey. The topics specified by the 16 respondents could be associated with survey topic 6, topic 7, topic 8 and topic 13 (see Table 3 for topic descriptions). Although additional data would be required to validate the results, the results indicate that respondents' research background might influence prioritisation of EE problems/topics of interest.

4.2   Skills Focus

The respondents had to indicate their skills focus for 6 skills categories, in terms of three levels: (1) knowledge obtained via course work only; (2) less than 3 years' work experience; and (3) more than 3 years' work experience. The question was optional, giving respondents the opportunity to only complete the relevant categories. Figure 2 indicates that for each of the 6 skills categories, more than 50 % of the respondents had significant work experience, i.e. *more than 3 years' work experience*. According to Fig. 2, respondents were especially skilled within the *enterprise design & modelling skills* category, which could influence prioritisation of EE problems/topics of interest.

4.3   Industry Background

The respondents obtained industry experience within a significant wide range of industries (24 industries out of 27 industries), meaning that EE/EA work within a vast majority of industries is represented. In terms of position, there is a balance between

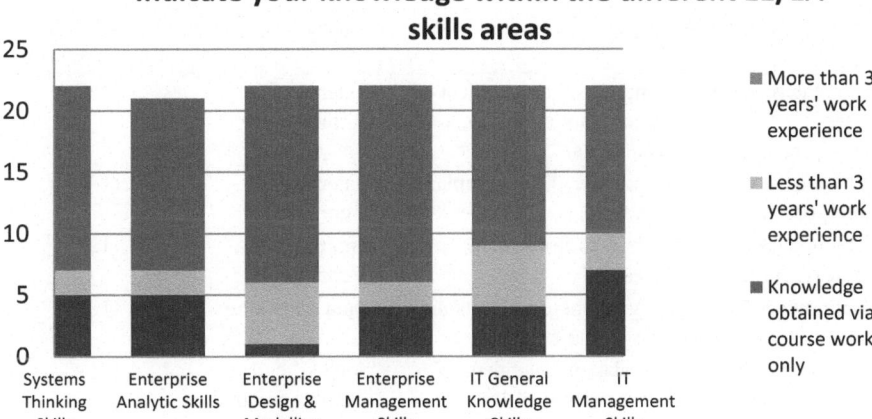

Fig. 2. Skills focus of the respondents

practitioners and academics: 8 of 22 respondents (36 %) have enterprise design/engineering-related positions, whereas 8 of 22 respondents (36 %) are academics.

The following three sections present results about the respondent-opinions regarding the demarcation of the *EE domain*.

4.4   The Phenomena of Interest for EE

Respondents had to select 4 (out of 9) phenomena that they would rank as the most important to address within the EE domain. Since 2 (of the 22 respondents) selected 5 (instead of 4) phenomena and 1 (of the 22 respondents) selected 3 (instead of 4) phenomena, a total number of 90 responses were received for prioritised phenomena of interest. The results indicate that 11 or more respondents (50 % or more) prioritised phenomena 1, 2, 3, 6 and 9. Although some phenomena received a low response in terms of importance (e.g. Phenomenon 8 was only selected by 3 of the 22 respondents), all of the 9 phenomena could still be included as valid EE domain topics.

Table 4 provides a summary of the prioritised phenomena, as well as themes that could be extracted from the phenomena descriptions. It is noteworthy that the latter choices in the questionnaire list received the least votes. This might be a bias introduced by the fact that the participants had to choose 4 out of a possible 9 phenomena of interest.

Respondents were also requested to add additional phenomena that need to be addressed within EE. Responses that could be associated with pre-defined phenomena of interest were included in Table 2. Other responses that indicate additional phenomena included:

- *"EE seems to be quite complete. The major obstacle I see for spreading the ideas of EE is education, first of the professors, then of the students."*

Table 4. Prioritised phenomena of interest within EE

Phenomena no. and description	Themes	Rating (number of responses)
1: The enterprise as a complex socio-technical system and its inability to adapt swiftly to rapidly-changing environments	Enterprise complexity	1 (20)
2: The inability to view/understand the complex enterprise in a holistic way	Enterprise complexity	2 (14)
3: The inability of enterprises to implement strategic initiatives successfully	Enterprise complexity	3 (12)
6: Data ownership challenges when data sets are shared beyond the scope of the enterprise	Data ownership	3 (12)
9: Society's concern with miss-management of natural resources and ill-treatment of low-income labour	Societal and environmental concerns	4 (11)

- *"The enormous chasm between what organisational sciences (already) know and what management does."*

4.5   The Problems/Topics of Interest for EE

Respondents had to select 10 (out of 22) problems/topics of interest that they would rank as the most important to address within the EE domain. Since 3 (of the 22 respondents) selected more than 10 problems/topics of interest and 10 (of the 22 respondents) selected less than 10 problems/topics of interest, a total number of 203 responses were received for including problems/topics of interest within the EE domain. The results indicate that 11 or more respondents (50 % or more) prioritised topics 2, 5, 6, 8, 9, 11, 13, and 15. As expected in Sect. 4.1, three of the prioritised topics (6, 8 and 13) corresponded with the research topics of the respondents. Although some topics received a low response in terms of importance (e.g. Topic 19 was only selected by 2 of the 22 respondents), all of the 22 topics could still be included as valid EE domain topics.

Table 5 provides a summary of the prioritised problems/topics of interest, as well as themes that could be extracted from the topic descriptions.

Respondents were also requested to add additional problems/topics that need to be addressed within EE. Some of these could be associated with the existing topics and the responses were added to Table 3. Additional problems or topics that could not be associated with existing topics were:

- *"It is my opinion that the use and impact of social media is under-estimated and should be considered."*
- *"It is not clear to me how the sub-fields combine to form the emerging discipline of EE as a whole. It appears to me that it is an extension of EA, but standard EA*

Table 5. Prioritised problems/topics of interest within EE

Topic no. and description	Themes	Rating (number of responses)
2: Define or align concerns or interests that should be addressed by the entire enterprise and its demarcated design domains	Alignment of concerns	4 (11)
5: Define the extent or scope of design/alignment/ governance in terms of internal enterprise structures AND extended enterprise structures	Scope of alignment	3 (13)
6: Architecture description, reference models and modelling practices for different enterprise design domains	Modelling	2 (15)
8: Design methodologies for designing the entire enterprise	Methodologies	1 (16)
9: Enterprise design methodologies that cover the entire enterprise life cycle	Methodologies	4 (11)
11: Governing mechanisms, practices (practice frameworks) and standards	Governance	3 (13)
13: Problem analyses practices and mechanisms to facilitate the identification of problems and the severity/effects of existing problems	Problem analysis	3 (13)
15: Impact analyses during enterprise re-design to estimate the impact/feasibility of different alternative solutions in terms of cost, schedule or change impact	Impact analysis	4 (11)

frameworks accommodate much of EE. So I am not sure how to differentiate EE from the evolving EA."

- *"What are the essential differences (if any) between the characteristics of 'normal' engineering sciences and enterprise engineering? How to integrate valuable insights from the traditional organisational sciences within the enterprise design perspective of enterprise engineering. Are the concepts we currently employ within enterprise engineering necessary and sufficient?"*
- *How to incorporate "potentially contributing complementary and objective bodies of knowledge to improve the definition of the field, namely those from viable systems theory and value systems theory".*

4.6   The EE Discipline Boundaries

For the final part of the survey, respondents had to comment on the ability or inability of using 4 pre-defined boundaries (defined in Sect. 2.2: *What are the boundaries of the discipline?*) to demarcate the boundaries of the EE discipline. The results indicated that 7 respondents (32 %) were unable to answer the question, while 2 respondents (9 %) disagreed with the boundaries. The remaining 13 respondents fully agreed (36 %) with the demarcation or partially agreed (23 %) with the demarcation. Respondents that partially agreed argued for excluding boundaries 2, 3 and 4. The majority of respondents (70 %) indicated that the philosophical boundary (*Concept of the enterprise and paradigm of creating value*) enables EE boundary demarcation.

Additional comments were also received from the respondents to demarcate the EE domain:

- *"Design scope should be defined as people, processes and resources."*
- *"The context in which the enterprise operates should be added."*
- *"Rather use objects and their life cycles to demarcate and ensure that EE does not overlap with EG, EA, SoS and SE design dimensions, frameworks, practices, methods and tools."*
- *"Boundaries cannot be set in stone. This applies specifically to the philosophical boundaries."*
- *"Academic curiosity will most likely not respect pre-defined boundaries".*

5 Discussion

This article used a questionnaire to (1) *validate* whether EECM represents the scope of the EE domain, and (2) *prioritise* the phenomena of interest and core problems or topics of interest within EE.

With regards to the *validation* of EECM, the results of the questionnaire indicated that all of the pre-defined phenomena of interest within EE were supported by the respondents. In addition, all of the pre-defined core problems or topics of interest within EE were supported by the respondents. Also, 36 % of the respondents agreed with the demarcated boundaries of the EE, whereas the majority of respondents (70 %) indicated that the philosophical boundary (*Concept of the enterprise and paradigm of creating value*) enables EE boundary demarcation. Therefore, there is strong evidence that *EECM can be used as a representation of the EE domain within the emerging EE discipline.*

Regarding the *prioritisation of phenomena of interest*, at least 50 % of the respondents prioritised 5 of the proposed 9 phenomena. From the prioritised phenomena, the *complexity of the enterprise* is the most prominent theme (see Table 4). Other themes that are less prominent include data ownership and addressing societal and environmental concerns. The phenomenon that was allocated the *lowest priority* was 'customers demanding always-availability of the enterprise'.

For the *prioritisation of problems or topics of interest*, 8 out of 22 topics were selected by at least 50 % of the respondents. Problems or topics of interest that received the *two highest rates* focused on two themes, methodologies and modelling (see Table 5). Even though a jungle of frameworks and modelling languages (with tool-support) already exist [16, 17], respondents still emphasise the need for methodologies and modelling. Three themes were in *third position*: (1) the scope of alignment, (2) governance, and (3) problem analysis; whereas two themes were in *fourth position*, alignment of concerns and impact analysis. The topics that were prioritised all represent *existing research topics* that feature in existing design/alignment/governance approaches [9, 12, 13], i.e. the futuristic topics incorporated from [14] were not prioritised by respondents. The two topics of interest that received the *fewest votes* were: (1) the need for tools to guide decision-makers in selecting EE software tools to model the enterprise or a specific design domain, and (2) the need to understand design/alignment/ governance approaches in terms of their *starting point* (e.g. top-down vs. bottom up) for modelling the enterprise.

Prioritisation of the phenomena and problems or topics provides a starting point for setting a research agenda within EE. Respondents also identified additional phenomena and problems or topics of interest: (1) differentiating between EA and EE, (2) integrating insights from traditional organisational sciences within the design perspective of enterprise engineering, (3) the observation that ideas of EE are not disseminated effectively via the education system, (4) the observed chasm between what organisational sciences already know and what management does, (5) the use and impact of social media, and (6) incorporating potentially contributing/complementary and objective bodies of knowledge to improve the definition of the field, namely those from viable systems theory and value systems theory. Yet, the additional phenomena and problems or topics do *not* relate to *domain questions*. According to Gregor [11], four classes of questions need to be answered when establishing a new discipline: (1) *domain questions*, the focus of this article, (2) structural or ontological questions; (3) epistemological questions, and (4) socio-political questions [11]. The additional phenomena and problems or topics of interest relate to the 2nd and 4th classes of questions, i.e. the structural or ontological questions (e.g. What is theory? How is this term understood in the discipline?) and socio-political questions (e.g. Where and by whom has theory been developed? Are scholars in the discipline in general agreement about current theories? How is knowledge applied? Is the knowledge expected to be relevant and useful in a practical sense?) Answering only the *domain questions* is not sufficient for establishing a new discipline. However, it is a prerequisite for initiating the debate about the emerging discipline.

6   Conclusion and Future Research

One of the key prerequisites for establishing EE as new discipline is to *define the domain of EE*. Previous work suggested that EECM could be used to define the domain of EE. In this article, a questionnaire was used to (1) to *validate* if EECM represents the scope of the EE domain, and (2) to *prioritise* the phenomena of interest and core problems or topics of interest within EE.

The sample of 22 respondents that participated in the questionnaire had a strong academic background, sufficient work experience within a significant wide range of industries. Although potential participants were concerned about the length and complexity of the questionnaire, the actual respondents not only completed the mandatory sections, but also provided additional comments and explanations.

The *validation* results confirmed that EECM embodies the *domain of EE*, since the respondents selected all the phenomena and problems or topics of interest, extracted from EECM, for EE domain inclusion.

The *prioritisation of phenomena of interest* highlighted *enterprise complexity* as the most prominent theme, whereas data ownership and addressing societal and environmental concerns were also prioritised. Enterprise complexity as a priority is expected since most literature on EA and EE alludes thereto. The other themes however warrant some future investigation and seem to indicate that EE is influenced by technological and societal trends as these topics were trendy at the time of writing this article.

The *prioritisation of problems or topics of interest* highlighted methodologies and modelling as the prominent themes. Although a number of theoretical methodologies, frameworks and modelling languages already exist, the respondent feedback indicates that more research is required. A study performed by Blowers [18] indicates that two thirds of sampled enterprises applied a hybrid approach, using multiple modelling languages and frameworks. Confirmed by some of the respondents, there is still a gap between what existing EE theory offers and its usefulness in practice.

The additional phenomena and problems or topics of interest that were identified by the respondents support the notion that *domain questions* alone are insufficient to establish a new discipline. The responses alluded to structural or ontological issues as well as socio-political issues. Further research is suggested to apply the four classes of questions presented by Gregor to establish EE as a discipline.

References

1. Giachetti, R.E.: Design of Enterprise Systems. CRC Press, Boca Raton (2010)
2. Bernard, S.A.: An Introduction to Enterprise Architecture EA3, 2nd edn. Authorhouse, Bloomington (2005)
3. Kappelman, L.A.: The SIM Guide to Enterprise Architecture. CRC Press, Boca Raton (2010)
4. Van Tonder, C.L., Roodt, G.: Organisation Development: Theory and Practice. Van Schaik Publishers, Pretoria (2008)
5. Hoogervorst, J.A.P.: Enterprise Governance and Enterprise Engineering. Springer, Diemen (2009)
6. Barjis, J.: Enterprise modeling and simulation within enterprise engineering. J. Enterp. Transform. **1**(3), 185–207 (2011)
7. Dietz, J.L.G., Hoogervorst, J.A.P., Albani, A., Aveiro, D., Babkin, E., Barjis, J., Caetano, A., Huyments, P., Iijima, J., Van Kervel, S.J.H., Mulder, H., Op't Land, M., Proper, H.A., Sanz, J., Terlouw, L., Tribolet, J., Verelst, J., Winter, R.: The discipline of enterprise engineering. Int. J. Organ. Des. Eng. **3**(1), 86–114 (2013)
8. Lapalme, J.: Three schools of thought on enterprise architecture. IT Prof. **14**(6), 37–43 (2012)
9. De Vries, M., Van der Merwe, A., Gerber, A.: Towards an enterprise evolution contextualisation model. In: First IEEE SMC Conference on Enterprise Systems (ES 2013). IEEE Explore, (2013)
10. De Vries, M., Gerber, A., Van der Merwe, A.: The nature of the enterprise engineering discipline. In: Aveiro, D., Tribolet, J., Gouveia, D. (eds.) Advances in Enterprise Engineering VIII, vol. 174, pp. 1–15. Springer, Switzerland (2014)
11. Gregor, S.: The nature of theory in information systems. MIS Q. **30**(3), 611–642 (2006)
12. De Vries, M.: A process reuse identification framework using an alignment model. Ph.D. Thesis. University of Pretoria, Pretoria (2012)

13. De Vries, M.: A framework for understanding and comparing enterprise architecture models. Manag. Dyn. **19**(2), 17–29 (2010)
14. Lapalme, J., Gerber, A., Van der Merwe, A., Zachman, J., De Vries, M., Hinkelman, K.: Exploring the Future of Enterprise Architecture: A Zachman Perspective (In review)
15. The Open Group: TOGAF 9.1 http://pubs.opengroup.org/architecture/togaf9-doc/arch/index.html
16. Schekkerman, J.: How to Survive in the Jungle of Enterprise Architecture Frameworks, 2nd edn. Trafford Publishing, Victoria (2004)
17. Schekkerman, J.: Enterprise Architecture Tools. (2012), http://www.enterprise-architecture.info/EA_Tools.htm#EAT
18. Blowers, M.: Hybrid Enterprise Architecture Frameworks Are in the Majority (2012), http://ovum.com/2012/03/22/hybrid-enterprise-architecture-frameworks-are-in-the-majority/

On Business Process Management, Simulation and Analysis

Towards Multi-perspective Modeling with BPMN

Richard Braun[(✉)] and Werner Esswein

Chair of Wirtschaftsinformatik, esp. Systems Development,
Technische Universität Dresden, Dresden 01062, Germany
{richard.braun,werner.esswein}@tu-dresden.de

Abstract. BPMN is the prevalent process modeling language and a lot of domain-specific BPMN extensions have evolved during the last couple of years. Due to the plenty of extensions and elements within BPMN, it is promising to consider complexity reduction mechanisms in order to provide appropriate, purpose-specific views on BPMN models. We therefore analyze capabilities of BPMN in regard of the definition of additional perspectives and diagrams in order to provide dedicated views on aspects of business processes (e.g., separate resource diagrams). As both BPMN and BPMN-defining MOF reveal shortcomings regarding to the definition of perspectives, we introduce a BPMN meta model extension in order to allow an integrated definition of new perspectives and their respective graphical elements. We further provide methodical guidance by conducting and customizing the BPMN extension method of STROPPI ET AL. (2011).

Keywords: BPMN extension · Concrete syntax · BPMN meta model · BPMN DG · Multi-perspectivity

1 Introduction and Motivation

The Business Process Model and Notation (BPMN) is the prevalent standard for business process modeling. BPMN is well-defined as meta model of the Meta Object Facility (MOF) in order to facilitate model exchangeability, tool integration and derivation of BPEL workflow models [1,2]. BPMN emphasizes that the modeling language only supports concepts that are applicable to business processes or required for their execution (e.g., *Data Objects*, *Participants* and even rather abstract *Resources*). For instance, it is explicitly stated that BPMN is not a data-flow language [1, p. 22]. However, it is necessary to focus BPMN language extensions as the prevalence and common application of standards lead to an increasing demand of language adaptation for specific purposes. The motivation for this comes from the promising application and exploitation of well established artifacts and their punctual customization for domain-specific features [3–5]. For instance, some businesses might require a more detailed representation of organizational structures or resources within BPMN in order to both benefit from the language strengths

© Springer International Publishing Switzerland 2015
D. Aveiro et al. (Eds.): EEWC 2015, LNBIP 211, pp. 67–81, 2015.
DOI: 10.1007/978-3-319-19297-0_5

on the one side (e.g., expressiveness, tool support, dissemination) and integrate additional concepts on the other side (cf. [3]).

Such new concepts are usually assigned to the *Collaboration Diagram* in BPMN as this is the most used diagram with a high expressiveness due to the variety of available concepts [2]. However, assigning each new concept to this diagram would lead to a loss of clarity and model readability [6, p. 9][1]. Thus, it is required to find a systematic way of structuring and filtering added concepts in order to provide stakeholder-specific models with a manageable level of complexity. Filtering relevant concepts can be seen as building particular views on the entire conceptual base by only showing those concepts that are interesting for a particular stakeholder group (e.g., *Data Objects* for IT engineers). As all those representations refer to the same conceptual base, integration remains possible. This aspect is referred as multi-perspectivity within the field of enterprise modeling and constitutes in integrated languages like ARIS [8] or MEMO [9]. So far, BPMN does not provide any consideration of multi-perspectivity. We argue that the integration of new perspectives within BPMN is very promising for model complexity management and enterprise modeling.

Hence, this research paper aims to provide meta model concepts for the definition of perspectives and diagrams within BPMN. We therefore consider the adaptation of the *Diagram Definition* (DD) specification in order to provide appropriate concepts within BPMN. Second, we aim to provide methodical guidance on building perspectives within BPMN extensions by enhancing an existing BPMN extension method [10]. The research article can be assigned to Design Science Research as it aims to develop artifacts in the form of meta model concepts and method adaptations [11,12]. The design is mainly driven by the adaptation of existing approaches in the context of BPMN. The designed artifacts are demonstrated by a BPMN extension of clinical resources in e-health.

The remainder of this paper is as follows. Section 2 provides some fundamentals on meta modeling and multi-perspective modeling in general. Section 3 presents the mechanisms for extending BPMN and considerable methods for systematic extension design. Section 4 introduces our approach for the integration of perspectives within BPMN. Afterwards, the extension method of STROPPI ET AL. (2011) is extended in Sect. 5 in order to facilitate a straightforward definition of perspectives. Thereby, our approach is demonstrated by a brief BPMN extension. The research article ends with a short summary and an outlook.

2 Fundamentals

2.1 Conceptual Models and Meta Modeling

A *conceptual model* is defined as the result of a construction process, "done by a modeler, who examines the elements of a system for a specific purpose" [6].

---

[1] Besides, BPMN generally struggles with the amount of rarely used notational elements [7].

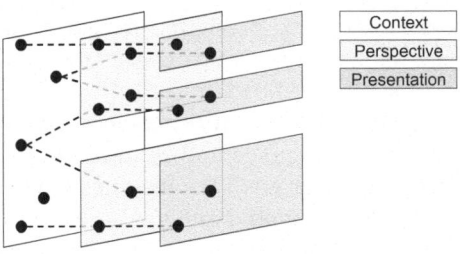

Fig. 1. Perspectives on contextual aspects and their graphical presentations.

In contrast to design models, which typically represent software systems or system parts, conceptual models represent real world phenomena [13]. Thereby, diagrammatic conceptual models have been established as an appropriate medium to foster communication between business stakeholders within the information system discipline [14]. Conceptual models are generated by conceptual modeling languages, which generally constitute as *semi-formal modeling languages*. These languages combine aspects of formal languages and natural languages. The *syntax* is defined formally within a meta model and can be divided into *abstract syntax* and *concrete syntax*. The first one defines the modeling grammar by specifying elements, properties, rules and constraints. The latter covers the graphical representation in the form of notational elements, views, perspectives and diagrams. The *semantics* arise not only from the syntactical constellation of a model, but also from specific domain terminology that is stated in a natural language [15, p. 111]. Modeling languages should always be embedded into a *modeling method*, which consists of the modeling language itself and a procedure describing the process of building and applying particular models [16]. Thereby, the *meta model* is defined as a specific model representing a modeling language [17,18]. Extending a modeling language both requires the existence of appropriate extension procedures (procedural aspect) and the existence of particular extension concepts within the meta model (structural aspect).

2.2 Multi-Perspective Modeling

Multi-perspective modeling is a specific technique in the area of conceptual modeling, which allows structuring an information system by different views in order to improve the understanding of its complexity [19]. By view building, the modeler can describe the entire information system using the views like a structuring framework, but he can also consider specific aspects of information systems by using the views for a aspect-specific model application in order to support the understanding of a particular domain [6]. Typically but not necessarily, one presentation type is assigned to each view, for instance, the org chart is used as presentation type for the organizational view. In the context of multi-perspective modeling, presentation types are not independent of each other. Moreover, they are linked by integrative model elements, i.e. elements that are used in multiple views. Thus, the model is a system of views, presentation types and the relationships between them. Views for structuring are specified by frameworks such as the *Architecture of Integrated Information Systems (ARIS)* [8].

Figure 1 demonstrates the idea of multi-perspectivity by filtering the entire contextual base of a modeling language and representing parts of those graphically in specific presentations (diagrams). For instance, it might be useful to establish process-oriented, document-oriented and goal-oriented perspectives within a BPMN model. Each perspective focusses on particular concepts or constructs of the modeling language in order to reduce the sheer amount of possible concepts that are not necessarily needed for answering questions that are related to particular model users. The stated perspectives can be refined in order to generate more fine-grained sub views. It could be also possible to set specific views, which allow the run-time visualization or hiding of elements (e.g., by refinements or compositions). Perspectives are then represented as diagrams with particular graphical elements.

2.3   Perspectives in Meta Modeling Languages

Meta Object Facility (MOF) and Diagram Definition (DD): MOF is the de facto standard for meta model definitions [18]. As MOF has its origins in the field of IT engineering, it has a strong focus on the abstract syntax of meta models and does not consider the concrete syntax explicitly. Perspectives are not specified and are totally in the responsibility of the particular language designers[2]. Reduction of model complexity can only be realized by package partitions within the abstract syntax [18, p. 9].

In 2012, OMG released the *Diagram Definition (DD)* standard aiming to provide a foundation for modeling and interchanging graphical notations [20, p. 1]. DD distinguishes two kinds of graphical information: First, graphics under the control of a user such as node positions and line routing points [20, p. 1]. These information need to be interchanged between tools as they refer to particular models on level M1. Second, there are graphics the users do not have necessarily control about as they are specified by the language standard (e.g., shapes and line styles). Although it is not explicitly stated in DD, those information represent the meta model based definition of the concrete syntax on level M2.

DD provides the *Diagram Interchange (DI)* package for the definition of user-specific graphical information [20, p. 21]. The *Diagram Graphics (DG)* package contains a model of graphical primitives for the mapping from the abstract syntax and DI to visual representations. DG is intended to contain all graphical information, a user does not have control over. Thus, the concrete syntax of the language can be defined effectively [20, p. 21]. DD explicitly states that it expects language specifications (e.g., the BPMN specification) to define mappings between interchangeable (DI) and non-interchangable graphical information (DG), but does not restrict the way of implementation within a particular

---

[2] Generally, the concrete syntax of a concept in MOF-based languages is defined in separate tables containing a graphic and a textual description of its appearance.

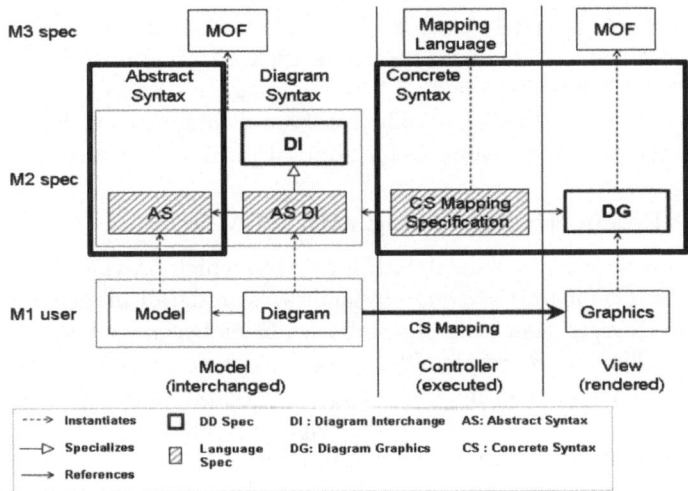

Fig. 2. Diagram definition architecture with particular consideration of language meta models [20, p. 5].

meta model (cf. "CS mapping specification" in Fig. 2) [20, p. 1]. Both DI and DG share common elements from the *Diagram Common (DC)* package which provides some primitive classes for diagram definitions [20, p. 7].

Other Meta Modeling Languages: MEMO is a framework for enterprise modeling and provides the MEMO MML language for the specification of domain-specific modeling languages (DSML) [9,19]. *Perspectives* encompass horizontal levels of considerations within enterprise models like strategy, organization or information system. Vertically, MEMO divides into specific *aspects* such a resources, processes or objectives [19, p. 1260]. MEMO does not provide dedicated concepts for the definition of graphics or diagrams [9, p. 22]. Moreover, the concrete syntax is not considered within the meta meta model [9, p. 25]. Perspectives are not specified explicitly in MEMO MML as they merely constitute within the designed DSMLs, which represent those particular perspectives and aspects [9, p. 31]. MEMO addresses multi-perspectivity by definition, but has a strong focus on dedicated DSMLs instead of widespread standard languages and their purpose-specific adaptation.

The **E3 meta modeling** approach is not very acquainted but provides a well-defined meta modeling language for the integrated definition of modeling languages [16]. E3 merges contextual aspects (e.g., objects and its properties) and presentational aspects (e.g., views and different presentations of a context object) and integrates them within one entire model. For instance, E3 allows the specification of different *Presentation Object Types* for particular concepts from the abstract syntax. Model elements from the context can be assigned to different *View Types* in order to establish user-specific filter mechanisms for complexity reduction. Within those views, *Presentation Types* can be defined in order to explicate particular aspects in different diagrams. The stated *Presentation Object Types* are assigned to those diagrams and can be specified by *Presentation Property Types* [16]. The integrated alignment of E3 regarding to perspectives seems to be promising for adaptation in the context of BPMN.

2.4   Multi-Perspective Modeling with BPMN

The process perspective is central within BPMN, which provides three modeling diagrams: The *Collaboration Diagram* facilitates in-detail modeling of business processes and the representation of collaborations between participants and business partners. The *Choreography Diagram* adopts a more global perspective and allows modeling message exchange between participants and is more message-oriented at all. Also, the *Conversation Diagram* is message-oriented, but has a more structural focus by emphasizing the relation between message elements. BPMN does not provide an explicit definition of perspectives or diagrams.

3   BPMN Extensibility in General

Adding perspectives and diagrams to an existing meta model affects both the abstract syntax and the concrete syntax as it has to be stated which concepts are assigned to which perspectives and diagrams. Hence it it necessary to examine BPMN's extensibility. With regard to the main parts of modeling methods, the consideration of BPMN is divided into the aspects *abstract syntax*, *concrete syntax* and *procedure*. The current state of the art of each aspect is analyzed with special regard to extending diagrams, views or perspectives.

3.1   Abstract Syntax

BPMN provides an extension mechanism that allows the definition and integration of domain-specific concepts and aims to ensure the validity of the BPMN core elements [1, p. 44]. A valid BPMN extension can be defined by the following elements: An *Extension Definition* is a named group of new attributes which can be used by other elements. It is possible to define new elements implicitly in that way. An *Extension Definition* consists of several *Extension Attribute Definitions*, which define attributes of new or original elements. Values of these *Extension Attribute Definitions* can be defined by the *Extension Attribute Value* class. The *Extension* elements bind a particular *Extension Definition* and its properties to a BPMN model definition. By doing so, all extension elements are accessible for existing BPMN elements [1, p. 58]. Although the extension mechanism provides specific concepts for meta model extension, there are some inaccuracies provoking confusion [21]. For instance, it remains unclear whether a BPMN extension generates a new meta model version or constitutes as profile in the sense of UML profiles [21]. BPMN does not provide concepts for the definition of perspectives as existing diagrams (e.g., collaboration diagrams) are only defined as elements of the abstract syntax without any specification of their particular "diagram" role [1, p. 109].

3.2   Concrete Syntax

In contrast to the abstract syntax, BPMN does not provide the precise definition of the concrete syntax of new elements as it does neither provide a specification of a BPMN DG package nor a respective mapping specification to the abstract syntax. Consequently, also the definition of new diagrams or perspectives is not considered. On model level M1, the concrete syntax of BPMN can be adapted by using colors for specific *Categories* or by the adaption of domain-specific icons for *Artifacts*. BPMN also provides an instantiation of the DI package for model interchange on level M1: BPMN DI supports interchange of shapes and edges which constitute a particular diagram [1, p. 367]. Thereby, a *BPMN Diagram* is understood as an incomplete or partial depiction of the content of the BPMN model [1, p. 367][3]. As BPMN does not provide a BPMN DG package or mapping rules, it is not possible to define user-independent graphical representations of the abstract syntax (cf. [20, p. 21]). Instead, those definitions rest on specific vendor implementations. Although this might be unproblematic for BPMN original elements, it remains tricky for extension elements as a general definition of their graphical appearance with DG is not designated so far.

3.3   Procedure

A systematic review of BPMN extensions by BRAUN & ESSWEIN (2014) reveals that only very few BPMN extensions make use of the BPMN meta model extension mechanism [3]. The missing methodical guidance within BPMN is suspected to be the main reason for that [3,21]. STROPPI ET AL. (2011) therefore propose a model-transformation based procedure model for BPMN extension development [10]. The procedure model consists of four main stages, which allow a straightforward derivation of valid BPMN meta models and XML definitions based on initial conceptual domain models [10, p. 5]. In the first step, the domain is conceptualized by a UML class diagram representing all new concepts and their relations to original BPMN elements. The *Conceptual Domain Model of the Extension (CDME)* consists of original elements (typed as *BPMN Concepts*) and extension elements (typed as *Extension Concepts* [10, p. 7]). In the next step, the CMDE is translated into the *BPMN Extension Model (BPMN+X)* by the application of model transformation rules. The BPMN+X model is specified by several stereotypes like *BPMN Element, Extension Element, Extension Definition* or *Extension Relationship* [10, p. 9]. This derived model can be applied as valid BPMN extension model. Finally, the extension model can be translated to XML schemes in order to facilitate model serialization [10, p. 11]. The approach provides a handy method for the definition of the abstract syntax of an extension. However, there is no consideration of adding perspectives to BPMN.

4   Extension of the BPMN Meta Model

Our examination in Sect. 3 reveals that the current BPMN meta model does not allow an extension with perspectives and diagrams for mainly two reasons: First,

---

[3] Again, it is important to notice that a *BPMN Diagram* refers to a specific instance of a BPMN diagram on level M1 (e.g., a collaboration diagram "Purchase").

these concepts are generally missing in BPMN and MOF-based languages on the level of abstract syntax. Second, there are no appropriate means for defining new diagrams on the meta model level, as BPMN only provides a BPMN DI specification for model exchange on M1 level.

Generally, it is necessary to define a BPMN DG model in order to facilitate vendor independent definitions of BPMN graphics and also define the concrete syntax of extension elements. With respect to the objective of our paper, we only want to focus the issue of adding perspectives and diagrams. We therefore propose an extension of the BPMN meta model by introducing some new concepts for the representation of perspectives since also the MOF specification does not provide appropriate concepts[4]. As meta model extensions are generally perceived as heavyweight operations, we intend to keep the number of new elements as small as possible.

Figure 3 presents our meta model extension. The extension attaches new concepts to the original extension mechanism of BPMN [1, p. 57]. *Base Elements* and *Extension Definitions* can be assigned to the introduced *Perspective* class in order to set a specific view or filter on the contextual entirety of BPMN concepts. The *canBeRepresented* attribute is added to the stated classes in order to allow some kind of a permission, whether a specific BPMN element can be represented graphically at all. *Base Elements* and *Extension Definitions* are separated in order to explicitly enable extension elements to be assigned to particular views. *Perspectives* have a particular *perspectiveName* for their identification (e.g., message view). They can be also nested within perspective hierarchies in order to define more detailed and restricted views. A particular *Perspective* is graphically represented by a *Diagram* that is identified by the *diagramName* attribute. In contrast to the *BPMN DI* class [1, p. 52], *Diagram* is intended to link abstract and concrete syntax within the meta model and not only for specifying diagram instances. Therefore, both *Perspective* and *Diagram* are defined as abstract super classes which need to be specialized by added or already existing diagrams (e.g., *Collaboration*)[5]. A *Diagram* can be represented by specific *BPMN Diagram* definitions defining the interchange format of diagram instances on level M1. We also introduce the *Diagram Node Adapter* class in order to integrate already existing diagrams to the proposed architecture. This is necessary as the three default diagrams in BPMN are defined as simple nodes within the abstract syntax and do not have any specific typing as diagram. It is then possible to assign those classes to the introduced *Diagram* class and exploit their capabilities. Also, it is possible to define particular *Perspectives* for the original diagrams (e.g., process perspective and message perspective).

Finally, a possible integration with elements from the DG package is outlined. Basically, two classes of DG are adapted by instantiation from DG to BPMN

---

[4] The name value pair extension tag concept from MOF is not applied as it is a rather slender mechanism, which does not allow complex structures [18, p. 23].

[5] Due to the customization of the BPMN extension structure, we also face the issue of intermediate abstraction levels, as a new perspective provokes a revision of the presented meta model (cf. [21]).

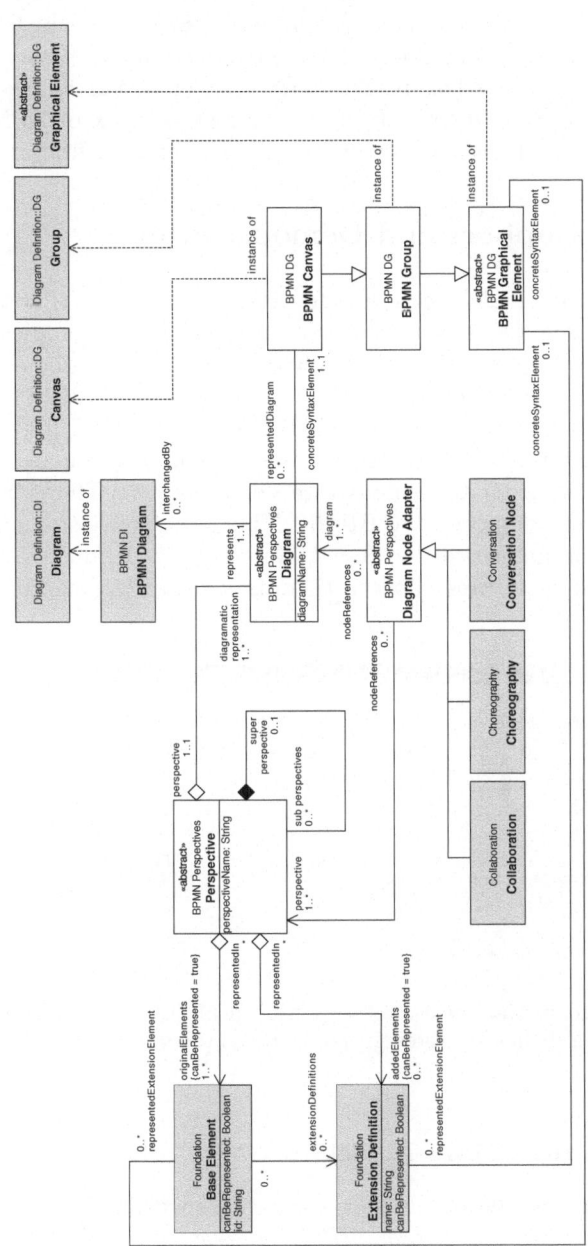

Fig. 3. Extension of the BPMN meta model by introducing classes for the definition of perspectives and diagrams. Further, an integration with an instance of the DG package is outlined. Original BPMN classes have a gray background; added elements have a white background.

DG: *BPMN Canvas* and *BPMN Graphical Element*, which are instances of *Canvas* and *Graphical Element*. *Canvas* represent the root of containment for all graphical elements within one diagram [20, p. 25]. Hence it is obvious to adapt it for diagrammatic representations. The *Graphical Element* class from DG is an abstract class for the specification of various notations such as *Rectangles*, *MarkedElements* or *Polylines* [20, p. 22]. With respect to the limited space of this paper, an element-wise definition of the concrete syntax of BPMN is not conducted, but the stated classes provide a base for further definition.

5 Methodical Support and Demonstration

After defining both the required concepts for abstract syntax and concrete syntax in Sect. 4, it is still necessary to provide appropriate methodical support for the definition and implementation of new perspectives and diagrams within BPMN. We therefore extend the method of STROPPI ET AL. (2011) by adding new stereotypes to the CDME model and the BPMN+X model. Figure 4 presents our extension. Each step is briefly described below by an ongoing example from the field of e-health in order to demonstrate our approach. Therefore, we aim to integrate a resource perspective to BPMN. This perspective should be represented as resource diagram in order to depict clinical resource bundles, containing clinical resources and structures between them in a separate diagram.

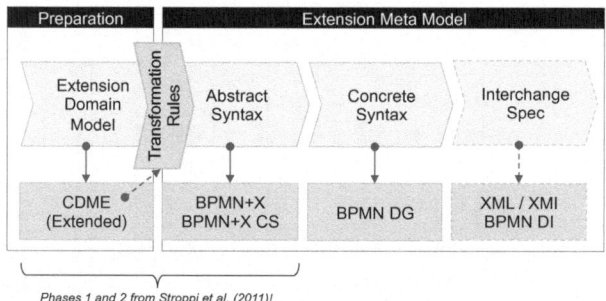

Fig. 4. Customizing the method of STROPPI ET AL. (2011) by introducing new stereotypes and integrating a dedicated stage for concrete syntax specification.

5.1 Extension of the CDME Model

We proclaim the introduction of four new stereotypes within the CDME model: *BPMN Diagram*, *Extension Diagram*, *BPMN Perspective* and *Extension Perspective*. The *BPMN Diagram* stereotype supports the assignment of extension elements to existing diagrams such as the collaboration diagram. The *Extension Diagram* stereotype indicates the specification of a new diagram that is assigned

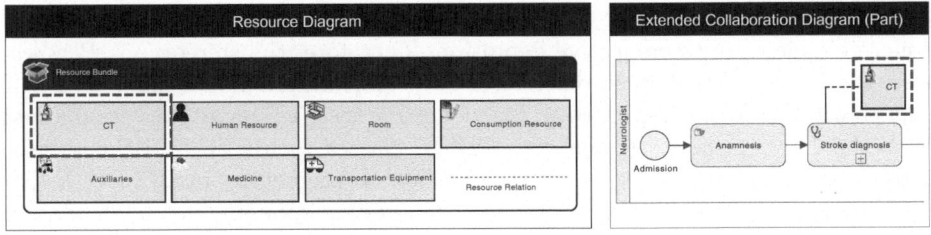

Fig. 5. CDME model of the proposed resource extension.

to a specific perspective. If the perspective is already defined, the *BPMN Perspective* stereotype can be applied. Otherwise, it is necessary to define a new *Extension Perspective*. A relation between an extension concept or BPMN concept to a class having a *BPMN Diagram* or having a *Extension Diagram* indicates a graphical element of this concept in the particular diagram. In the same way, a relation to a particular perspective is realized[6]. Thus, it remains possible to assign a concept to a perspective, but omit the graphical definition. This is useful for modeling non-graphical elements which primarily act as property ranges. In Fig. 5, the *Consumption Resource* and the *Auxiliaries* concept should not be represented graphically, for instance.

Figure 5 covers all resource-related aspects that are considered within our demonstration case. Basically, the BPMN class *Resource* was specified by particular sub types from the domain of clinical resources (e.g., *Equipment*, *Medicine* or *Room*). Thereby, the specification of *Human Resources* is possible. Objects of this class can be represented by particular *Participants* from BPMN. A *Resource* can be assigned to different *Resource Bundles* in order to encapsulate related resource objects. The *Resource Relation* class allows the specification of relations between all resource-related elements. Its values are defined within the *resourceRelationType* attribute that is defined within an enumeration. *Resources* and *Resource Bundles* are related to *Activities*. Additionally, a *Resource Perspective* and an respective *Resource Diagram* are modeled. Their containing concepts are assigned by modeling associations between single classes. It is also required to integrate specific graphics of *Resources* and *Resource Bundles* within BPMN processes. Therefore, associations between the *Collaboration Diagram* class and both concepts are modeled. It means, that graphical representations of both elements should appear within the collaboration diagram of BPMN.

5.2   Extension of the BPMN+X Model

The BPMN+X model facilitates the straightforward definition of valid extension models and is created by the application of 15 transformation rules [10, p. 10]. Due to the introduced stereotypes in the CDME model, the BPMN+X definition has to be extended, too. In addition to the casual BPMN+X model (which

---

[6] Of course, extension models with many considered concepts should be divided into separate packages in order the ensure model readability!.

defines the abstract syntax), we define an additional BPMN+X CS model for the concrete syntax and perspective definition. Therefore, the stereotypes *Perspective* and *Diagram* are defined representing the aimed meta model classes. CDME classes with *BPMN Perspective* or *Extension Perspective* stereotypes are marked with the *Perspective* stereotype and CDME classes with a *BPMN Diagram* or *Extension Diagram* stereotype with the *Diagram* stereotype in BPMN+X CS. Technically, this typing indicates the creation of corresponding instances of associated meta model elements.

Further, all the elements have to be considered. First, the original BPMN+X transformation rules have to be conducted in order to derive *Extension Definitions*, *Extension Elements* and *BPMN Elements* (cf. [10, p. 10]). Then, it has to be checked whether an association between such an element and a perspective exists. If this is the case and if there are no semantical contradictions in regard of the BPMN specification, then the particular element can be assigned to the perspective within the meta model (cf. Table 1). In the BPMN+X CS model, the association between the elements remains visible in order to represent their relation. The same procedure has to be applied to diagrams in order to analyze elements that need to be represented graphically. If any element should be defined graphically within a diagram, then the *BPMN Graphical Element* stereotype should be added in the BPMN+X CS model[7]. Accordingly, a *BPMN Canvas* stereotype has to be assigned to respective diagrams.

5.3   Concrete Syntax of the Extension and Interchange Specification

Due to our above presented meta model extension, the concrete syntax of both diagrams and extension elements can be defined by adaptation of the DG package which is instantiated to BPMN DG. All added diagrams imply the creation of an instance of *BPMN Canvas* with the name of the diagram. All added extension elements having a relation to any diagram and its respective perspective needs to be specified, too. Those elements indicate the creation of a *BPMN Graphical Element* instance and a particular configuration and specification in alignment to the DD package at all. Figure 6 depicts the concrete syntax of our demonstration case by introducing the *Resource Diagram* and the new *Resource* element within the *Collaboration Diagram*.

The interchange of extension elements can be realized as follows. The abstract syntax of the extension can be interchanged by the translation of the BPMN+X model to the XML schemes defined in [10]. The exchange of the concrete syntax is divided into two parts. Those elements, a user cannot change or redefine, can be described by the DG package (e.g., BPMN DG). Thus, the concrete syntax of BPMN extensions could be exchanged between modeling tools. However, the BPMN DG package needs to be specified in a dedicated research article. Single BPMN model instances can be exchanged by the application of the BPMN DI definition as stated in Sect. 3.2.

---

[7] Please note, that the in-detail definition of BPMN DG is not within the scope of this research article due to its limited space of pages.

Table 1. Added transformation rules for the concrete syntax (BPMN+X CS model) representing the relation between BPMN concepts or extension concepts and particular perspectives or diagrams within the CDME model.

CDME Stereotype	BPMN+X Stereotype (Concept)	Additional Stereotype in BPMN+X (Concept)
BPMN Perspective	BPMN Element	Not permitted!
BPMN Perspective	Extension Element/Def.	-
Extension Perspective	BPMN Element	-
Extension Perspective	Extension Element/Def.	-
BPMN Diagram	BPMN Element	Not permitted!
BPMN Diagram	Extension Element/Def.	BPMN Graphical Element
Extension Diagram	BPMN Element	BPMN Graphical Element
Extension Diagram	Extension Element/Def.	BPMN Graphical Element

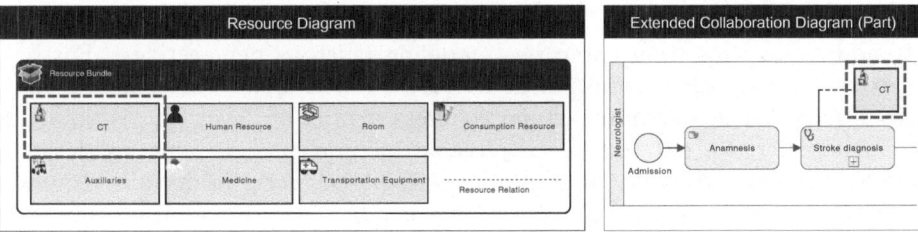

Fig. 6. On the left side, the new *Resource Diagram* is presented. Also, the concrete syntax of its containing elements is depicted. On the right side, the extended *Collaboration Diagram* is demonstrated by a process part containing an *Activity* which uses a particular *Resource* element that is represented in both perspectives ("CT" stands for computed tomography).

6   Conclusion

This paper addresses the topic of multi-perspectivity within the process modeling language BPMN. As BPMN is very popular and widely used both in academia and professional practice, BPMN is likely to be used as central tool within enterprise modeling. Based on its extension mechanism, various domain-specific extensions can be added to BPMN. However, BPMN does not provide any means for managing the complexity resulting from those extensional elements. Due to the stated aspects, we propose the definition of perspectives and diagrams within BPMN as suitable means for complexity management and a better separation of concern within the entire modeling language. After an examination of the capabilities of both MOF and BPMN regarding the definition of those elements, we decided to extend the BPMN meta model in order to both integrate perspectives and diagrams as well as enable the systematic definition of the concrete syntax by leveraging the *Diagram Graphics* package from DD. Both issues are integrated within a minor meta model extension that establishes the base for an in-detail definition of BPMN DG, for instance.

Our approach is the first one explicitly addressing perspectives in BPMN and their precise definition within its meta model. Methodical support is given by the customization of a BPMN extension method. However, the proposed approach can be also used for casual modeling: Due to the mass of graphical elements in BPMN, it might be very promising to define user-specific perspectives and reduce visible elements. Our approach can also be applied for a better integration of single perspectives into BPMN (e.g., in the field of enterprise architecture modeling).

Nevertheless, there are some tasks for further research: The presented BPMN meta model extension reveals a minor abstraction shortcoming that is inherently related to the extension mechanism at all [21]: Actually, the meta model provokes the definition of a revision of the meta model, when some extension element is added to it (in detail, cf. [21]). This issue is caused by the inability of MOF to define perspectives and diagrams within one language meta model. Thus, we recommend further research on extension mechanisms in MOF, which explicitly address aspects from enterprise modeling (such as perspectives). Further, a precise and detailed definition of BPMN DG is required in order to allow a tool-independent definition of the concrete syntax on level M2. Within our research paper, we only outline a possible integration. Last but not least, a detailed demonstration of our technique within real world projects is required. Currently, a prototypical implementation is conducted within a business process modeling project in the e-health sector.

References

1. OMG: Business Process Model and Notation (BPMN) - Version 2.0. Object Management Group (OMG) (2011)
2. Chinosi, M., Trombetta, A.: BPMN: an introduction to the standard. Comput. Stand. Interfaces **34**(1), 124–134 (2012)
3. Braun, R., Esswein, W.: Classification of domain-specific BPMN extensions. In: Frank, U., Loucopoulos, P., Pastor, Ó., Petrounias, I. (eds.) PoEM 2014. LNBIP, vol. 197, pp. 42–57. Springer, Heidelberg (2014)
4. Braun, R.: Towards the state of the art of extending enterprise modeling languages. In: 3rd International Conference on Model-Driven Engineering and Software Development (2015)
5. Brinkkemper, S., Saeki, M., Harmsen, F.: Meta-modelling based assembly techniques for situational method engineering. Inf. Syst. **24**(3), 209–228 (1999)
6. Schuette, R., Rotthowe, T.: The guidelines of modeling - an approach to enhance the quality in information models. In: Ling, T.-W., Ram, S., Li Lee, M. (eds.) ER 1998. LNCS, vol. 1507, pp. 240–254. Springer, Heidelberg (1998)
7. Muehlen, M., Recker, J.: How much language is enough? theoretical and practical use of the business process modeling notation. In: Bellahsène, Z., Léonard, M. (eds.) CAiSE 2008. LNCS, vol. 5074, pp. 465–479. Springer, Heidelberg (2008)

8. Scheer, A.-W., Nüttgens, M.: ARIS architecture and reference models for business process management. In: van der Aalst, W.M.P., Desel, J., Oberweis, A. (eds.) Business Process Management. LNCS, vol. 1806, pp. 376–389. Springer, Heidelberg (2000)
9. Frank, U.: The memo meta modelling language (MML) and language architecture. ICB Research report 24, Universität Duisburg-Essen (2008)
10. Stroppi, L.J.R., Chiotti, O., Villarreal, P.D.: Extending BPMN 2.0: method and tool support. In: Dijkman, R., Hofstetter, J., Koehler, J. (eds.) BPMN 2011. LNBIP, vol. 95, pp. 59–73. Springer, Heidelberg (2011)
11. Hevner, A.R.: The three cycle view of design science research. Scand. J. Inf. Syst. **19**(2), 87 (2007)
12. Winter, R.: Design science research in Europe. Eur. J. Inf. Syst. **17**(5), 470–475 (2008)
13. Wand, Y., Weber, R.: Research commentary: information systems and conceptual modeling - a research agenda. Inf. Syst. Res. **13**(4), 363–376 (2002)
14. Frank, U.: Conceptual modelling as the core of the information systems discipline-perspectives and epistemological challenges. In: AMCIS 1999 Proceedings, p. 240 (1999)
15. Pfeiffer, D., Gehlert, A.: A framework for comparing conceptual models. In: Proceedings of the Workshop on Enterprise Modelling and Information Systems Architectures, pp. 108–122 (2005)
16. Greiffenberg, S.: Methodenentwicklung in Wirtschaft und Verwaltung. Kovač, Hamburg (2004)
17. Strahringer, S.: Metamodellierung als Instrument des Methodenvergleichs: Eine Evaluierung am Beispiel objektorientierter Analysenmethoden. Ph.D. thesis, TU Darmstadt (1996)
18. OMG: Meta Object Facility (MOF) Core Specification, Version 2.4.2 (2014)
19. Frank, U.: Multi-perspective enterprise modeling (MEMO) conceptual framework and modeling languages. In: Proceedings of the 35th Annual Hawaii International Conference on System Sciences, pp. 1258–1267 (2002)
20. OMG: Diagram Definition (DD), Version 1.0 (2012)
21. Braun, R.: Behind the scenes of the BPMN extension mechanism - principles, problems and options for improvement. In: 3rd International Conference on Model-Driven Engineering and Software Development (2015)

Analysis of Business Processes with Enterprise Ontology and Process Mining

Artur Caetano[1,2](✉), Pedro Pinto[3], Carlos Mendes[3], Miguel Mira da Silva[1,3], and José Borbinha[1,2]

[1] Instituto Superior Técnico, University of Lisbon,
Avenida Rovisco Pais 1, 1049-001 Lisbon, Portugal
[2] INESC-ID, Rua Alves Redol 9, 1000-029 Lisbon, Portugal
artur.caetano@tecnico.ulisboa.pt
[3] INESC-INOV, Rua Alves Redol, 9, 1000-029 Lisbon, Portugal

Abstract. This paper describes a business process analysis method that helps determining if a business process complies with the requirements put forward by enterprise ontology's transaction pattern. The method starts by discovering the business process through the application of process mining techniques to the events that are generated by the applications that support the execution of the process. This step discovers the actual implementation of the process from its event trace. Next, the discovered process is analysed against enterprise ontology's transaction pattern to determine whether the process complies with the structure and sequencing of its coordination and production acts. The paper shows that combining process mining with enterprise ontology contributes to the analysis of business processes, especially in terms of determining the boundaries of authority and responsibility of the process. The feasibility and the limitations of the method are discussed using a case study that analyses a semi-automated business process.

Keywords: Process analysis · Process mining · Enterprise ontology · DEMO

1 Introduction

Business processes can be abstracted through the use of conceptual models [1]. Models provide the means to analyse and communicate the structure and relationships of the abstracted concepts according to specific purposes and views [2,3]. Techniques such as Event-Driven Process Chains (EPC), the Business Process Model and Notation (BPMN) and ArchiMate use different concepts to represent different aspects of an organization and its business processes. However, if the goal is redesigning a process, these modelling techniques need to be complemented with business process redesign techniques [4,5] that often synthesize best practices and empirical results [6–9]. However, process models are often

This work was supported by national funds through Fundação para a Ciência e a Tecnologia (FCT) with reference UID/CEC/50021/2013.

designed with the goal of communicating ideas and not to support their analysis or operation, which translates to process specifications that are incomplete, informal, or ambiguous [10]. Furthermore, analysing a process requires having an up-to-date model that reflects the actual operations of the organization. Some approaches to process acquisition and discovery rely on interviews, surveys, and document analysis [11], while others rely on deriving architectural models from organizational events [5,12].

This paper proposes a method to analyse whether the activities of a business process comply with a pattern that describes how actors should cooperate and commit to an agreement regarding the production of a service or product. This type of analysis is especially important to cross-functional processes due to the communication that is required to coordinate the different actors [10,13]. The activities of a cross-functional process are performed by multiple actors or intersect different organizational units. Analysing how actors communicate helps determining the actors responsible for performing an activity, if an actor is authorized to perform an activity, and whether an activity has been delegated to another actor [14].

The method is grounded on process mining techniques [5,15] and enterprise engineering theories [14,16]. Process mining is used to discover an up-to-date version of the business process, while the theories of enterprise engineering are used to ground the collaborative communication patterns that are used to analyse the process. This approach meets three goals of enterprise engineering [16] as it (*i*) accounts people as a valuable asset of an enterprise; (*ii*) considers the different enterprise domains as a whole; and, (*iii*) relies on the PSI-theory which describes the operation of enterprises, supporting business understanding and enterprise changes in a way that makes those organised complexities manageable.

The method takes as input an event log that captures events that are generated by a set of automated and semi-automated applications that support the implementation and execution of instances of business processes. The event log is then used as input to a process mining algorithm that statistically infers the underlying process model. Next, the discovered process is analysed and a set of DEMO models [14], namely an Actor-Transaction Diagram and a Process-Structure Diagram, are used to specify its actors, transactions, along with the coordination and production acts. At this stage, the method enables assessing the consistency and completeness of the discovered business process against enterprise ontology's transaction pattern. The process is deemed *consistent* if its activities can be fully enclosed within one or more business transactions. This means that the activities of a consistent process describe how a requester and a provider cooperate and commit to an agreement regarding the production of a service or product. Moreover, the sequencing of these activities must also comply with the sequencing of the transaction pattern. However, a process may be consistent but may be incomplete because it lacks some coordination steps defined in transaction pattern. The process is deemed *complete* if all steps within the transaction pattern can be mapped to activities of the process.

This paper demonstrates the application of the method to analyse a semi-automated process that implements the access approval to a VPN within defence governmental institution. The paper is organized as follows: Sect. 2 summarizes the foundations of enterprise ontology. Section 3 describes the method and its steps in detail. Section 4 presents the case study. Finally, Sect. 5 discusses the contributions and limitations of the proposal and concludes the paper.

2   Foundations of the PSI-theory

The Design and Engineering Methodology for Organisations (DEMO) is used to model, design and (re)engineer organisations and networks of organisations [14]. DEMO is grounded on the performance in social interaction theory (PSI-theory), which explains the construction and operation of enterprises. The principle of enterprises is that employees, together with representatives of customers and suppliers, enter into and comply with commitments regarding the products they cooperatively produce. This understanding makes enterprises social systems of which the elements are human beings in their role as social individuals, to whom appropriate authority is granted, and bearing the corresponding responsibility [16]. The PSI-theory provides a notion of *enterprise ontology* that is defined as the full implementation-independent understanding of the essence of an enterprise's organization. It posits four main principles, namely: (*i*) the operation axiom, (*ii*) the transaction axiom, (*iii*) he composition axiom, (*iv*) the composition axiom, and (*v*) the organisation theorem.

These principles specify how actors establish commitments and communicate. The next sections briefly introduce the four axioms.

2.1   Operation Axiom

The operation axiom states that the operation of an enterprise is constituted by the activities of actor roles, which are elementary chunks of authority and responsibility, fulfilled by subjects. In doing so, these subjects perform two kinds of acts: *production acts* and *coordination acts*. By performing production acts (p-acts), subjects contribute to bringing about the goods or services that are delivered to the environment of the enterprise (the production facts). By performing coordination acts (c-acts) subjects enter into and comply with commitments towards each other regarding the performance of production acts. A subject in its fulfilment of an organizational role is called an *actor*.

2.2   Transaction Axiom

A transaction describes how a particular result (a *fact*) is produced as the collaboration between two actor roles, an initiator and an executor. These two roles may be played by the same actor in the case of a self-activated transaction. Thus transaction specifies how coordination acts and production acts relate with each other, and who performs each act. A transaction therefore encloses a pattern of well-defined coordination and production steps.

A transaction is organized in three phases: the order phase (O-phase), the execution phase (E-phase), and the result phase (R-phase). During the *order phase*, the initiator actor and the executor actor work together to reach an agreement about the intended result of the transaction, i.e., they agree on the fact to be produced and on its qualities. During the *execution phase* the executer produces a single production fact. In the *result phase*, the initiator and the executor work to reach an agreement about the acceptance and delivery of the production fact. In an optimistic scenario, where the initiator and executor always come to an agreement, the collaboration pattern follows the so called basic transaction pattern Fig. (1) that comprises four coordination acts (*request, promise, state, and accept*) and one production act. The *request* and *accept* coordination acts are performed by the initiator whereas the *promise* and *state* coordination acts and the fact production are performed by the executor. The basic pattern is limited as it assumes that the initiator and the executor keep consenting to each other's acts. The full standard transaction pattern extends the basic pattern with the *decline, quit, reject* and *stop* acts to represent the scenarios where the actors disagree or fail to produce the results.

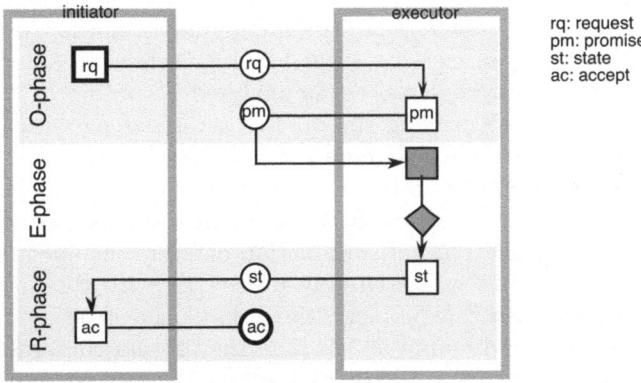

Fig. 1. The basic pattern of a transaction (from [14]).

2.3   Composition Axiom

The composition axiom states that a transaction can be enclosed or nested within another transaction. This axiom provides the basis to define a business process, where a business process is a collection of causally related transaction types, so that the starting step is a request to an actor role from the environment (an external activation) or a request to an internal actor role from itself (a self-activation).

2.4   Distinction Axiom

The distinction axiom identifies three distinct human abilities that are able to play a role in the operation of actors: *performa*, *informa* and *forma*. These abilities regard communication, creating things, reasoning, and information processing. The ability that deals with the form aspects of communication and information is *forma*. It involves uttering and perceiving sentences, the syntactical analysis of sentences, coding schemes, transmission of data, storage and retrieval of data or documents. The *informa* ability abstracts the form aspects and is concerned with the content of communication and information as sharing of thoughts between people, the remembering and recalling of knowledge and reasoning. *Performa* concerns the bringing about of new, original things, directly or indirectly by communication as commitments, decisions and judgments.

3   The Business Process Analysis Method

The method aims to analyse a business process in terms of the collaboration between its actor roles. It receives as input an event log and generates as the final output a revised process model that is complete and consistent according to enterprise ontology's transaction axiom.

The method combines process mining and DEMO and comprises the tasks represented in Fig. 2. In *step 1*, a process mining algorithm is used to discover a business process from an existing event log, represented in a normalized format. In *step 2*, the discovered process is analysed according to the operation axiom with the goal of identifying the process actors, along with its coordination and production acts. *Step 3* represents the discovered process as DEMO diagrams (an ATD and PSD). *Step 4* deals with the analysis and revision of the DEMO diagrams according to the transaction and composition pattern so that the missing coordination and production acts are included. Finally, *step 5* revises the process that was originally discovered with the goal of making the process consistent (so that its activities respect the structure of a basic or standard transaction) and complete (so that the process contains all required transactional coordination and production acts). The revised process can now be used to facilitate the actual redesign of the process and its implementation. Note that the method relies on the implicit discovery of process models and not on the explicit acquisition of process models. This means that the method can be applied iteratively without the need of keeping process models. The steps of the method are described next.

Step 1 *Discover the process model.* This automated task discovers a business process from an event log in XES format [17]. The process is discovered using the Flexible Heuristics Miner (FHM) [18] with the ProM 6 tool [17,19]. Although other mining algorithms could have been selected, the FHM attempts producing process models that can be understood and analysed by humans instead of models that are optimized for computational analysis. To do so, the algorithm tries to balance the complexity of the

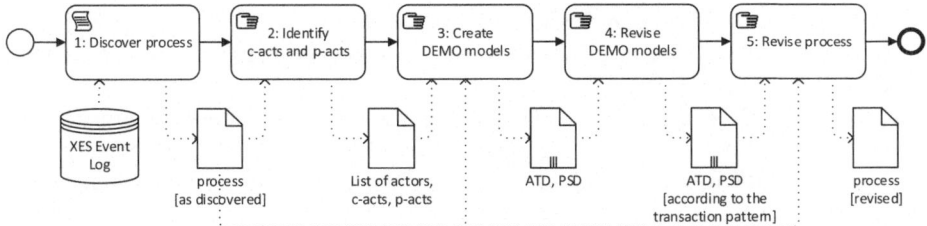

Fig. 2. The process analysis method.

discovered model in terms of its core structural elements and its details. This property is particularly appealing because the application of enterprise ontology's principles to process model analysis requires to discover the core activities of a process and not the details behind its control flow mechanisms. The FHM is also designed to handle noisy event logs and event types with a low degree of structuring, which also contributes to generate models that focus on the main elements of the process.

Step 2 *Analyse the discovered process model.* This step analyses the activities of the process and classifies them according to the operation and distinction axioms. The operation axioim classifies an activity as a production or coordination act. The distinction axiom further classifies an activity as a *performa, informa,* or *forma* act. The operation axiom also identifies the actor roles involved in the process. The result of this step is a list of actors, coordination and production acts that is fully traceable to the elements of discovered process model from where they were sourced.

Step 3 *Create the DEMO models.* This step produces DEMO ATD and PSD diagrams. It first selects the subset of *performa* coordination and production acts from the results of step 2. These *performa* acts are the ontological acts that need to classified according to the transaction axiom. Based on the transaction axiom, the discovered activities are classified according to the basic transaction pattern, i.e. as *request, promise, state* or *accept* c-acts or as p-acts. The process may also be analysed according to the standard transaction pattern in can the analysis requires such level of detail. In this case, the standard pattern is used and the activities can also be classified as *decline, reject, stop* and *quit* c-acts. Note that this step reproduces the discovered process model and thus it may produce incomplete and inconsistent DEMO models.

Step 4 *Revise the DEMO models.* This step revises the DEMO models produced in step 3. The revision consists of two sub-steps. The first is adding to the DEMO models the missing acts that are required according to the transaction pattern. This makes the model complete. The second step is making sure that the revised transaction goes through the sequence steps defined in the O-phase, E-phase, and R-phase. The composition axiom is then applied to ensure that all of the transaction steps follow a logical sequence according to the pattern. The second sub-step makes the model consistent. The result of step 4 are the revised DEMO diagrams along with a gap analysis report that describes the changes that were made to the original models so that they become compliant.

Step 5 *Revise the process model.* This step revises the discovered process according to the results of step 4. The goal is to include in the process the activities that were not found in step 4. The revised process becomes compliant with the transactional pattern, meaning it explicitly contains the required transactional steps with the proper sequencing.

4   Demonstration

The demonstration was performed using data from a process within a national defence government institution. One of the services the institution provides is the administration of shared IT infrastructures, including databases, operating systems, and internal networks. It also provides user support through a centralized service desk. The processes realizing these services are semi-automated or fully automated, meaning they are performed by a combination of people and applications or by a set of applications. Moreover, these systems have monitoring capabilities that can be used to generate events triggered whenever an instance of the process meets a certain condition. As a result, the events generated by the multiple instances of the process can be recorded into an event log.

The events that were used in this case study result from a semi-automated process that manages the granting of access to virtual private networks. The operations performed by the applications or by the users of the applications are registered into an event log. Each event describes a case identifier (i.e. a unique identifier of each process instance), a timestamp, the request (e.g. access, the source/owner of the request (e.g. user X, service Y), and the response (e.g. denied). The requests and responses are classified according to a taxonomy of types. Note that the type and number of events that can be logged depends on the capabilities of the systems under observation and on the process management framework [20]. There are several business process management techniques that can be used to facilitate the definition, generation, and processing of events and to prepare the event log for adequate process mining and analysis [5]. However, the discussion of this topic is out of the scope the present paper.

A case of the VPN access approval process always starts with an access request, followed by a notification to the requester that the request was successfully registered. The registration assigns a case or ticket number to the request. Upon registration, an approver is notified of the request. The approver is responsible for analysing the request, deciding on accepting or rejecting the request, and documenting the reason behind the decision. An IT unit then implements the access in case of a positive decision from the approver. The requester is notified of the result and the process ends when a confirmation is received from the requester. The overall process also handles access refusal and the compensation actions required to address problems but these actions are not considered here. The following sections describe the application of each step of the method.

The method starts with an event log generated from events produced by the business process support applications. All cases unrelated to the VPN access approval were first filtered out, including the events that related with denial of access. The results here reported are based on a set of approximately 1000 events derived from around 50 different cases that ended with the successful granting of approval. As such, the next steps are analysed with the basic transaction pattern. An analysis of the complete process would follow the same method but use standard transaction pattern instead. In term of approach, we recommend first performing a simpler compliance analysis against the basic transaction pattern, then iterating over the process design until achieving compliance, and only then analysing the process against the standard transaction pattern.

4.1 Step 1: Discover the Process Model

The first step infers the process model from an event log in the XES format. As described before, this is accomplished automatically through the Flexible Heuristics Miner and the ProM tool. Figure 3 depicts the main activities of the discovered workflow.

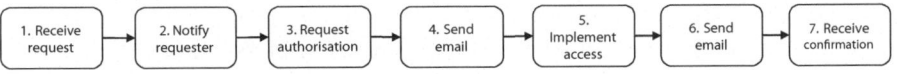

Fig. 3. The core workflow discovered by the process miner.

The activities identified by the miner are primarily derived from events triggered by the sending or receiving of messages. The process starts with the reception of an access request (activity 1). The next activity is sending a ticket to the requester that represents the successful registration of his request (activity 2). Next, the miner discovered the sending of the access request from a system to the approver (activity 3). Since the log is only documenting successful approvals, the authorization request is always followed by sending an email to the implementation team requesting the configuration (activity 4). The implementation team then marks the successful implementation of the access in one of the support systems, which triggers an observable event (activity 5). The miner also captures the next activity which consists of sending a message to the requester with the credentials and access instructions (activity 6). Finally, the requester confirms that the access was successfully granted (activity 7).

4.2 Step 2: Analyse the Discovered Process Model

The next step analyses the discovered activities according to the operation and distinction axioms and associates the acts to transactions. The results are summarized in Table 1.

Table 1. Activity classification according to the operation, distinction and transaction axioms

	Activity	Distinction	Operation	Transaction
1	Receive request	Ontological	c-act	T01/request
2	Notify requester	Info/datalogical		
3	Request authorization	Info/datalogical		
4	Send email	Ontological	c-act	T02/request
5	Implement access	Ontological	p-act	T02/execute
6	Send email	Ontological	c-act	T01/state
7	Receive confirmation	Ontological	c-act	T01/accept

Two out of the seven discovered activities are classified as non-ontological, namely the sending of the ticket to the requester (activity 2), and the sending of the authorization request to the approver (activity 3). Note that the generation of the ticket (activity 2) corresponds to a infological or datalogical notification and does not correspond to the ontological *promise* that would confirm that the access was going to be granted. Similarly, activity 3 is not ontological because it is a signal between a system and the approver and as such it does not represent the actual production act of T01. The classification of "send email" (activities 4 and 6) as an ontological act seems arguable because these activities appear to datalogical or infological acts. However, the information that is conveyed by these two activities actually represents the results of ontological commitment: activity 4 signals the implementer to start a service (T02/request), and activity 6 sends a business object to the requester that represents the completion of the service by the executor (T01/state). Therefore, these two activities are considered ontological.

4.3   Step 3: Create the DEMO Models

This step uses the data in Table 1 along with actor (user) information available in the event log to create a DEMO construction model, the Actor Transaction Diagram (ATD), depicted in Fig. 4. The overall access approval process consists of two causally related transactions T01 and T02, being T02 enclosed within T01 according to the composition axiom. T01 represents the service that is available to the external environment and that grants (or denies) a requester the access to a VPN. This transaction is initiated by a requester actor role and executed by an approver actor role. The approver role stands for an actor that is properly authorized by the organization to decide on granting access to the requester. This actor also becomes responsible for the decision. Transaction T02 is initiated by the approver and executed by the IT team that implements and configures the VPN access.

Combining the ATD with the acts identified in the previous step and summarized in Table 1, results in the DEMO process structure diagram depicted in Fig. 5. This process diagram lacks several coordination and production actions,

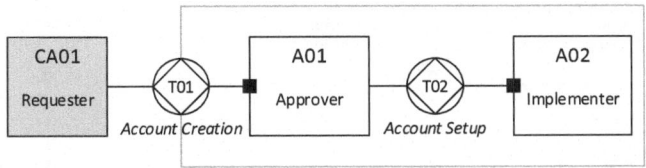

Fig. 4. Actor Transaction Diagram generated from the discovered process.

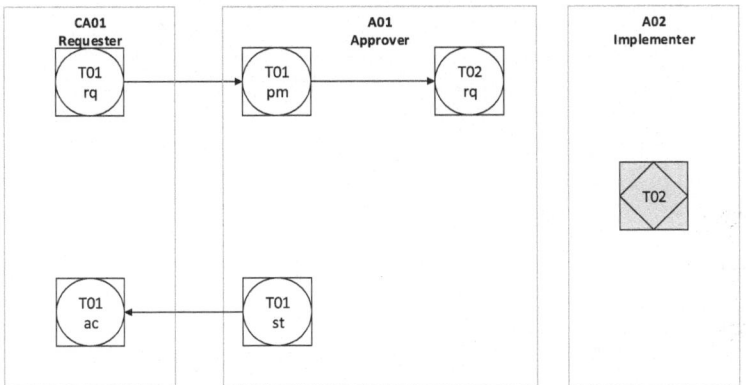

Fig. 5. Process diagram according to the discovered process model.

which means the discovered process model is not complete. Transaction T01 lacks the p-act and T02 lacks the promise, state and accept c-acts that establish the commitment between the implementation team and the approver.

4.4   Step 4: Revise the DEMO Models

The incomplete process model is revised as to include the missing coordination and production acts. In this case study only the acts from the basic transaction pattern are being considered. Using the complete pattern would identify the c-act T01/decline that would be executed instead of T01/promise whenever the approver decided not to grant access to the requester. Additional acts would also be identified to handle rejections of the service or failures in the approval or implementation tasks. The revised process model is depicted in Fig. 6.

4.5   Step 5: Revise the Process Model

The final step uses the artefacts produced in steps 2, 3 and 4 to revise the process discovered in step 1 (cf. Fig. 3). It produces a labelled process model that classifies the type of each activity according to enterprise ontology's principles, and identifies the acts that are missing and make the process incomplete. Figure 7 shows the revised process model, where the dashed activities are not part of the discovered process.

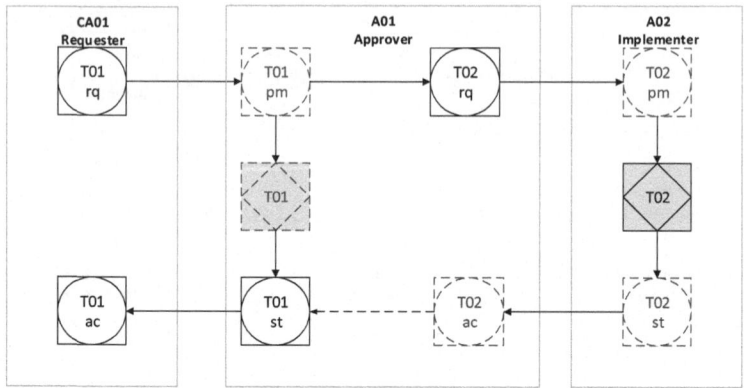

Fig. 6. Revised process diagram highlighting the missing acts in the discovered process model.

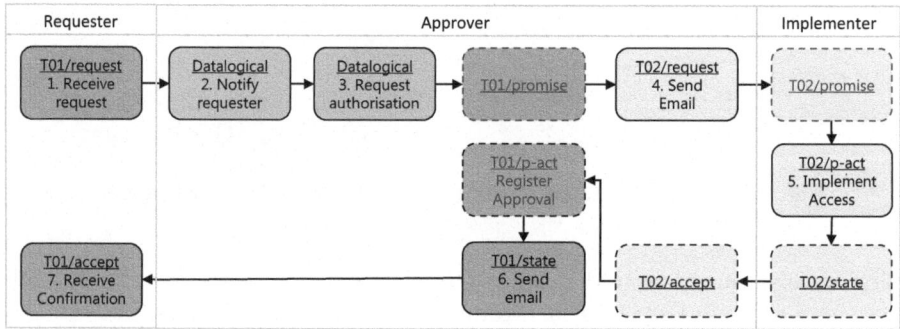

Fig. 7. Revised process.

A straightforward approach to redesign the process entails adding five activities, one for each missing ontological act: T01/promise, T01/p-act, T02/promise, T02/state and T02/accept. The revised business process becomes complete and consistent and therefore compliant with transaction pattern. However, this straightforward approach may not produce optimal results since it does not redesign the remaining process structure. As such, the process discovered in step 1 should be first analysed using business process management techniques [5] before applying more complex analysis and redesign techniques.

5   Conclusion

The motivation behind this project is to evaluate the applicability of using enterprise ontology's principles to analyse processes that are not explicitly modelled. This paper describes a method that analyses whether a business process complies with the requirements put forward by enterprise ontology's transaction pattern. Processes are discovered using process mining techniques that analyse the events generated by their implementation on the business support systems and infrastructure. The structure and the activities of the discovered process are then evaluated against the structure and the coordination acts defined in the transaction pattern. The evaluation highlight the deviations between the discovered process and the expected process structure as prescribed by the transaction pattern. As such, the analysis focusses on how actors collaborate via coordination acts and not how the process produces its services or products. This contributes to identifying the boundaries of authority and responsibility within the process, which plays an important roles when designing business processes that involve multiple actors, organizational units or functions. The results of the analysis can be used to redesign the implementation of the so that authority and responsibility become clear.

The case study confirms the overall feasibility of the approach. Nevertheless, further case studies with larger and more complex process are needed in order to assess the complexity of classifying the activities and identifying transactions according to the principles of enterprise ontology. The main limitations of the method are proportional to the limitations of process mining, which in turn depend on the quantity and quality of the events that can be analysed. If the event log does not adequately represent a process, then the mining techniques will be unable to discover processes with statistical confidence. But even if a process is discovered with a high-degree of confidence, the mining algorithms may infer activities with a low level of granularity because discovery is based on events generated by the implementation of a process and not on its high-level business description. This means that the discovered activities are likely to translate to datalogical and infological acts and not directly to ontological acts. As such, the analysis of the discovered process must be able to deal with the clustering of activities at different levels of detail and the underlying semantic gap. However, there are techniques that aim addressing such type of issues and that should be explored. These techniques include the semantic annotation of event logs [21,22], the application of natural language processing [23], and the application of ontologies to analyse architectural models [24].

Our current work involves extending the process discovery step to include business and data objects as a means to help identifying transactions through the identification of production facts. We are also working on combining the semantic annotation of event logs [22] with ontologies as a means to describe and analyse multiple models [24].

References

1. Albani, A., Dietz, J.L.G.: Enterprise ontology based development of information systems. Int. J. Internet Enterp. Manag. **7**(1), 41–63 (2011)
2. Greefhorst, D., Proper, E.: Architecture Principles: The Cornerstones of Enterprise Architecture. Springer, Heidelberg (2011)
3. Proper, H.A., Lankhorst, M.M.: Enterprise architecture - towards essential sensemaking. Enterp. Model. Inf. Syst. Archit. J. **9**(1), 5–21 (2014)

4. Becker, J., Kugeler, M., Rosemann, M. (eds.): Process Management: A Guide for the Design of Business Processes. Springer-Verlag, Berlin (2003)
5. Dumas, M., Rosa, M.L., Mendling, J., Reijers, H.A.: Fundamentals of Business Process Management. Springer, Heidelberg (2013)
6. Barjis, J.: Automatic business process analysis and simulation based on demo. Enterp. IS **1**(4), 365–381 (2007)
7. Bose, R.P.J.C., Mans, R.S., van der Aalst, W.M.P.: Wanna improve process mining results? In: CIDM, pp. 127–134. IEEE (2013)
8. Jansen-Vullers, M.H., Kleingeld, P.A.M., Loosschilder, M.W.N.C., Netjes, M., Reijers, H.A.: Trade-offs in the performance of workflows – quantifying the impact of best practices. In: ter Hofstede, A.H.M., Benatallah, B., Paik, H.-Y. (eds.) BPM 2007 Workshops. LNCS, vol. 4928, pp. 108–119. Springer, Heidelberg (2008)
9. Reijers, H., Mansar, S.L.: Best practices in business process redesign: an overview and qualitative evaluation of successful redesign heuristics. Omega **33**(4), 283–306 (2005)
10. Vergidis, K., Turner, C.J., Tiwari, A.: Business process perspectives: theoretical developments vs. real-world practice. Int. J. Prod. Econ. **114**(1), 91–104 (2008)
11. Liles, D.H., Presley, A.: Enterprise modeling within an enterprise engineering framework. In: Winter Simulation Conference, pp. 993–999 (1996)
12. Tribolet, J., Sousa, P., Caetano, A.: The role of enterprise governance and cartography in enterprise engineering. J. Enterp. Model. Inf. Syst. Archit. **9**(1), 28–49 (2014)
13. Antonucci, Y.L., Goeke, R.J.: Identification of appropriate responsibilities and positions for business process management success: seeking a valid and reliable framework. Bus. Proc. Manag. J. **17**(1), 127–146 (2011)
14. Dietz, J.L.G.: Enterprise Ontology: Theory and Methodology. Springer, Heidelberg (2006)
15. IEEE Task Force on Process Mining: Process mining manifesto. Technical report (2011)
16. Dietz, J.L.G., Hoogervorst, J.A.P.: The unifying role of enterprise engineering. In: Magalhaes, R. (ed.) Organization Design and Engineering, pp. 11–38. Palgrave Macmillan, Basingstoke (2014)
17. Verbeek, H.M.W., Buijs, J.C.A.M., van Dongen, B.F., van der Aalst, W.M.P.: XES, XESame, and ProM 6. In: Soffer, P., Proper, E. (eds.) CAiSE Forum 2010. LNBIP, vol. 72, pp. 60–75. Springer, Heidelberg (2011)
18. Weijters, A.J.M.M., Ribeiro, J.T.S.: Flexible heuristics miner (FHM). In: IEEE Symposium on Computational Intelligence and Data Mining, pp. 310–317. IEEE (2011)
19. Process Mining Group, Eindhoven Technical University: Prom tools. http://www.promtools.org/. Accessed 01 April 2015
20. van Dongen, B.F., de Medeiros, A.K.A., Verbeek, H.M.W.E., Weijters, A.J.M.M.T., van der Aalst, W.M.P.: The ProM framework: a new era in process mining tool support. In: Ciardo, G., Darondeau, P. (eds.) ICATPN 2005. LNCS, vol. 3536, pp. 444–454. Springer, Heidelberg (2005)
21. Medeiros, A.K.A., van der Aalst, W., Pedrinaci, C.: Semantic process mining tools: core building blocks. In: 16th European Conference on Information Systems (ECIS) (2008)
22. de Medeiros, A.K.A., van der Aalst, W.M.P.: Process mining towards semantics. In: Dillon, T.S., Chang, E., Meersman, R., Sycara, K. (eds.) Advances in Web Semantics I. LNCS, vol. 4891, pp. 35–80. Springer, Berlin Heidelberg (2009)

23. Leopold, H.: Natural Language in Business Process Models. LNBIP, vol. 168. Springer, Heidelberg (2013)
24. Antunes, G., Barateiro, J., Caetano, A., Borbinha, J.: Analysis of federated enterprise architecture models. In: 23rd European Conference on Information Systems (ECIS) (2015)

A Case Study of Business Process Simulation in the Context of Enterprise Engineering

Yang Liu[1,2(✉)] and Junichi Iijima[1]

[1] Department of Industrial Engineering and Management,
Tokyo Institute of Technology, Tokyo, Japan
{liu.y.af,iijima.j.aa}@m.titech.ac.jp
[2] Department of Software Engineering,
Shenyang Normal University, Shenyang, Liaoning, China

Abstract. The high failure rates occur in many real-life business process reengineering projects discovered that the current methods in business process re-design and reengineering are not effective for supporting the requirements of change. To avoid the limitations of current mainstream workflow based perspective and methodology, the authors proposed a simulation framework in the context of enterprise engineering, named DEMO++. This research aims to apply the DEMO++ based simulation in a real-world case study with the following objectives: to evaluate the advantages, potentials and limitations of DEMO++ based simulation; to further investigate how it can assist in business process change; and to find problems to be improved in the selected process of "Company C".

Keywords: DEMO · DEMO++ · Business process simulation

1 Introduction

When an organization carefully re-designs their business model, vision, and mission, their business processes must be quickly restructured to support these upper-level changes. However, contrary to these goals, remarkably high failure rates occur in many real-life business process reengineering projects based on several recent research studies [1–3]. These previous studies have discovered that the current methods in business process re-design and reengineering are not effective for supporting the requirements of change. The authors considers the limitations in business process change analyses are caused by limitations in the traditional workflow perspective and in the available research methods.

- The workflow perspective places too much emphasis on details without developing a holistic, high-level perspective. Moreover, the enterprise is neither considered as an entire system nor components in workflow perspective, such that these methods are typically non-modularized and weak in supporting change analysis,
- In addition, current modeling methods lack of tools for evaluating the effects of designed solutions before implementation; meanwhile, current simulation methods are weak at describing large, complex system; it has limited ability for the design of 'to-be' models and lack of methodology to confirm the consistency of change.

D. Aveiro et al. (Eds.): EEWC 2015, LNBIP 211, pp. 96–110, 2015.
DOI: 10.1007/978-3-319-19297-0_7

In order to solve these problems, a simulation framework DEMO++ is proposed [4] from the perspective of enterprise engineering. This research aims to apply DEMO++ in "Company C", with the following objectives:

- To evaluate the advantages, potentials and limitations of DEMO++ based simulation;
- To further investigate how it can assist in business process change;
- To observe problems in the "proposal and estimation process of company C" and to provide suggestions for improvement.

The remainder of this paper is organized as follows: Firstly, Sect. 2 introduced the concepts of business process simulation and DEMO++; Secondly, Sect. 3 described in detail the case, DEMO aspect models, DEMO based DEMO++ models, DEMO++ based AnyLogic simulation, results and feedbacks. Finally, Sect. 4 gave a brief discussion and a future work proposal.

2 Literature Review

Simulation allows and more importantly provides powerful assistance to generate new ideas for change, to explore the effects of alternative changes, to implement those changes without disrupting the business system and to compare the performance of both the present and reengineered systems [5]. The technique of using simulation in the context of a business process is referred to as *business process simulation* (BPS) [6]. BPS is a powerful tool that can assist in change analysis and effectiveness evaluation due to its ability to measure performance, to test alternatives and to engage in processes [5].

However, there are several barriers that prevent BPS from being widely used in business process change analyses. As some researchers have noted [7, 8], most business process improvement projects consider only a single process without taking a holistic perspective of the enterprise; the complexity of simulation increases when individual small process models are joined into a large hierarchical construct. Thus, it is inefficient to utilize these methods in process change analyses that concern an entire enterprise. Moreover, changing models are very complex. Simulation is a useful tool in comparing 'as-is' and 'to-be' models to validate the effects of change and to ensure the completeness of a model; however, most of the business process simulation literature restricts itself to comparing the before and after conditions, providing little support on the redesign process [9].

All these limitations are caused by the weakness in conceptual modeling, the most important but difficult step of simulation. As indicated by Bank et al. [10], there are surprisingly few books and academic papers on the subject of building conceptual models for enterprise-related simulations. Currently, the most popular conceptual models for business process simulation is based on workflow perspective. However, as mentioned above, this perspective leads to the un-structured and none-modularized simulation model such that it has difficult to confirm the consistency before and after change.

Conceptual model is expected to be more structured, modularized and formal. Turnitsa et al. [11] suggested that a conceptual model should be "an ontological representation of the simulation that implements it". One example of an ontology-based conceptual model for simulation is the *system entity structure* (SES) proposed by Zeigler [12]. SES was utilized in several of Zeigler's discrete event simulation studies. However, SES emphasizes only the system's data structure. Therefore, it is poor at describing enterprises as social systems and is less applicable in the context of business processes.

DEMO is an enterprise ontology proposed by Dietz [13] in order to abstracted the essence of enterprise by ignoring all the implementation details. Comparing with other workflow based models, this methodology is good at analyzing enterprise from a high abstracted level to assist in grasping the essence of the enterprise. In order to combine the advantage of DEMO with the dynamic analyzing methodologies, a serious of researches has been conducted by CIAO group members. Representative one could be Barjis' DEMO based petri-net simulation, which makes DEMO model executable and quantitatively analyzable [14]; Another one is Kervel's DEMO processor [15], which is a powerful tool for simulating coordination processes following DEMO definition. All these researches are DEMO based that the simulations are in the ontological level. However, the author argues that only ontological level cannot fully support BPS requirements in business process improvement or re-engineering, since there are different level of changes: some of them are coordination and cooperation related changes in ontological level but the others are implementation related changes that may or may not compare with ontological changes. Since DEMO ignored all implementation details, it may have difficulty to be employed directly for describing all the change types in BPS.

In order to take the advantages of DEMO and to avoid the limitations, a DEMO++ framework is proposed in another manuscript of the authors' [16]. DEMO++ aims at providing a layered simulation framework that allows the analysis on ontological level changes, implementation level changes or both [16]. By combing ontology defined in DEMO with implementation details, DEMO++ can answer the questions, "who coordinate", "coordinate for what" and "how to coordinate". So that DEMO++ based simulation can assist in measuring not only the effects of ontological changes but also the effects of changes related with the implementation of coordination.

3   A Case Study

In order to evaluate the proposed DEMO++, a case study is conducted in "Company C", a large Japanese information system integrator from September, 2014 to December, 2014 in its "proposal and estimation process". The process is described in Sect. 3.1.

3.1   Case Description

The customers' requirements are first sent to the salesperson in charge in "Company C" for **evaluation**[1]. *If the customer can meet the criteria, they will be accepted. If the customer cannot meet any of the criteria, they will be rejected. In situations in which the case is complex, the sales manager* **decides whether to initiate a case receipt symposium (CRS)**. *In requesting a CRS, after discussion, the* **CRS record is organized** [2] *and* **confirmed** *by the sales manager.*

When a case reception is accepted, "Company C" must **evaluate the risk of the case**. *In the risk evaluation process, the salesperson in charge* **prepares all of the documents**[3]. *According to the risk level, the sales manager* **decides whether to request a prior review board (prior RB or PRB)**. *If a PRB is necessary, after discussion, the* **PRB record is organized** *by the salesperson and* **confirmed** *by the sales manager.*

The salesperson **prepares a proposal estimation and a proposal** *that refers to the PRB record, the customer information, and the risk review sheet. After the documents are submitted for review. The reviewers include other staff members from the sales department, the sales manager and staff from the development department. If the* **proposal is not acceptable**, *it must be* **redone**. *In contrast, the salesperson* **requests an estimation symposium to evaluate the estimate**. *The salesperson also responds by* **preparing the required documents** *for the estimation symposium; he/she also* **arranges the symposium**. *In the symposium, the* **estimate is evaluated** *following risk evaluation rules.*

An unacceptable **estimate must be redone**. *For any accepted proposals and estimates, the sales manager* **decides if a regular review board (regular RB or RRB) is necessary** *based on the risk level of the proposal. In a regular RB,* **high-risk proposals and estimates are discussed and re-evaluated**. *If the proposal is acceptable, the* **regular RB record is prepared and** *submitted to the QA department for a* **commitment**. *The QA department* **decides whether to ask for an executive symposium** *according to the RB result and the risk level of the case. If an executive symposium is required, the sales manager must* **prepare the documents** *for an executive symposium and* **submit** *them to the executive office. Members who attend the executive symposium typically include the sales manager and the executive officers. If the proposal is* **not acceptable**, *it must be* **re-proposed**. *If the proposal* **is committed, it is proposed to the customer** *as a final solution.*

3.2   DEMO Aspect Models

DEMO aspect models show snapshots of an enterprise from different aspects. A *construction model* (CM), is the most concise model that describes how transactions and actor roles are composed to construct a system. A *process model* (PM) describes the detailed causal relationships and constructions that exist in processes. A *fact model* (FM) describes the objects and facts that are related to a process. Lastly, an *action model* (AM) describes the action rules for the actor roles.

---

[1] The business-level activities are denoted in font "**Agency FB**" (in bold).

[2] The data-level activities are denoted in font "Agency FB".

[3] The information-level activities are denoted in font "Agency FB" (with underline).

The CM of company C is shown in Fig. 1. Nine transaction types are defined to indicate different objectives. Transaction types T1, T2 and T3 are defined for the case reception; transaction types T4 and T5 are defined for the case risk evaluation; transaction type T6 is defined for the proposal and estimation; and transaction types T7, T8 and T9 are defined for the proposal and proposal evaluation. There are two information banks:

- AT1: The customer base includes all customer-related information (e.g., company name, sales turnover, and business).
- AT2: The rule base includes all rules related to the proposal and estimates inside the company (e.g., risk-level division rules).

Proposal& Estimate

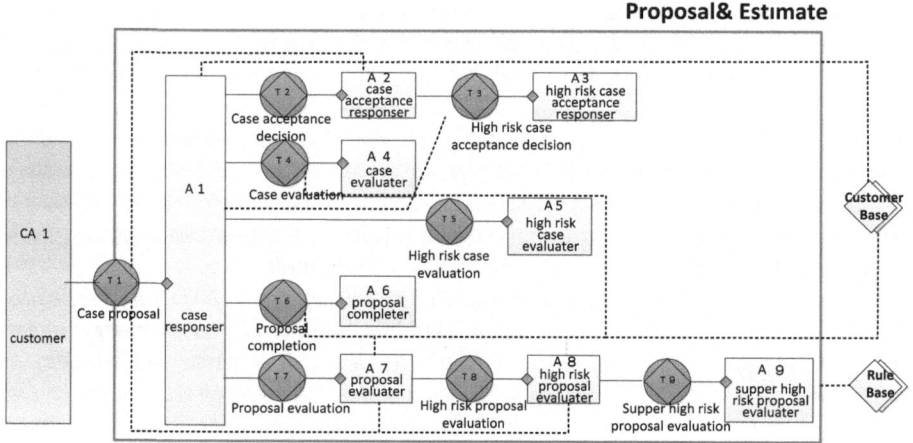

Fig. 1. OCD of "Company C" proposal and estimation process

In DEMO PM for Company C, a process structure diagram (PSD) is subsequently defined to further describe the coordination details of the processes, as is shown in Fig. 2.

- The promise of T1 "case proposal" [T1/pm][4] must wait until the acceptance of T2 "case receipt decision" is complete (T2/ac)[5].
- In T2 ("case receipt decision"), some of the "high-risk cases" must also be evaluated by T3 ("high-risk case receipt decision"); therefore, those acts [T2/ex] must wait for the "acceptance of high-risk case receipt decision" (T3/ac) fact.
- When a "case proposal" is promised (T1/pm), the case must be evaluated. Similar to a case receipt, a case evaluation must also consider high-risk cases that require additional evaluation through T5 ("High-risk case evaluation"). Thus, the act request of T6 [T6/rq] must wait until all case evaluation processes are complete, i.e., either (T4/ac) or (T5/ac).

---

[4] [ ] represents act: [T1/pm] means act "to promise an instance of T1".

[5] ( ) represents fact: (T2/ac) means fact "an instance of T2 has been accepted".

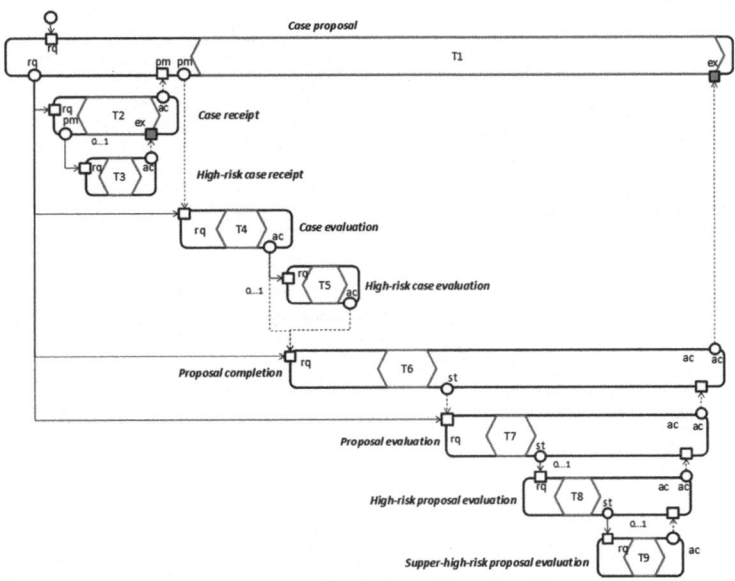

Fig. 2. PSD of "Company C" proposal and estimation process

- T6 ("proposal complementation") requires an evaluation or several evaluations according to the risk level of the proposal; thus, the acceptance of T6 [T6/ac] must wait for one or all facts [T7/ac] (normal-risk proposal), [T8/ac] (high-risk proposal) and [T9/ac] (super-high-risk proposal) according to the risk level of the case proposal.

In DEMO FM, an Object Fact Diagram (OFD) is defined. As shown in Fig. 3, there are three object types: the Case (S), Customer (C) and Proposal (P), whereby each of which include various defined properties. The Case (S) is related to five productions, i.e., P1 - P5), and the Proposal (P) is related to four productions, i.e., P6, P7, P8 and P9. Here, the risk level of each case could be directly assigned or calculated according to items listed in a "risk sheet".

3.3   DEMO++ Models

Following Zeigler's framework (2000), DEMO++ is designed as a component based structure, which is easy to be change. There are three parts included: ontology, implementation and the main. The Main part and the ontological part of DEMO++ are directly derived from the DEMO CM, PM, FM and AM by a predefined mapping rule and a set of corresponding developed transforming tools [4] (more detailed explanation of transformation is expressed in another manuscript [16]).

- *Ontology part* (as shown in Fig. 4) describes the ontology of an enterprise, including the components of the transaction types (T) and aggregate transaction types (AT), such as information banks; object types (O); and actor roles (AR).

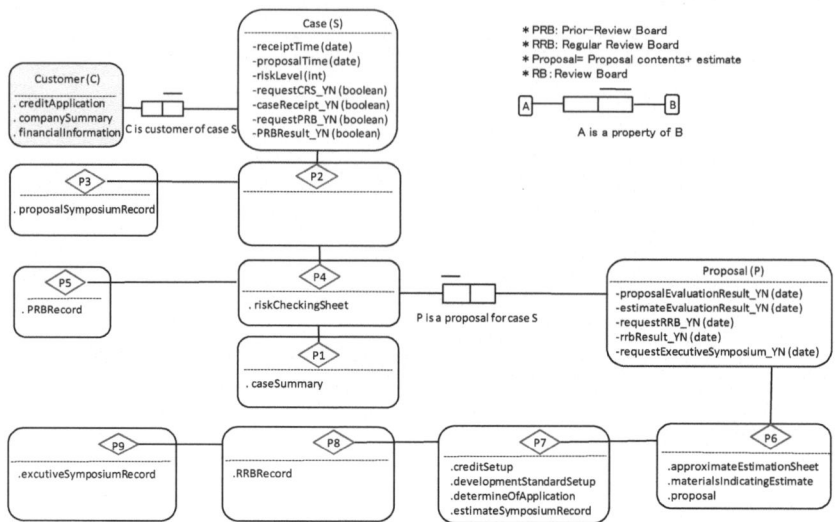

Fig. 3. OFD of "Company C" proposal and estimation process

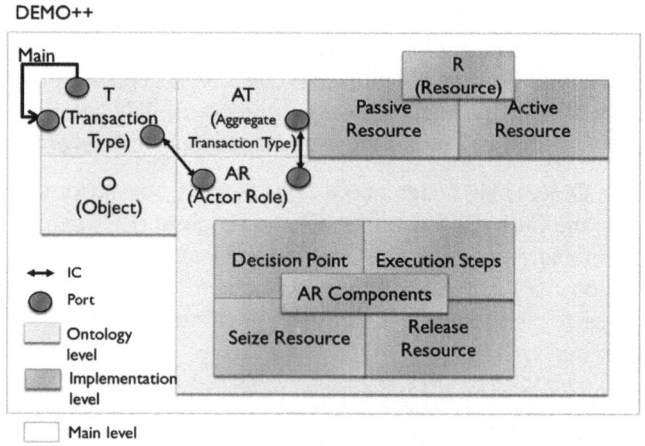

Fig. 4. DEMO++ (Color figure online)

- Transaction types (T) are generally defined following a standard transaction pattern, as expressed in the left side of Fig. 5. This pattern can be changed into a completed transaction pattern or a simple one. For each specific transaction type (T), the parameter settings are required, as given in the right side of Fig. 5. Most of the setting are automatically derived from the DEMO aspect models, except the "call processor" items, which need to be additionally added to explain whether an act is need to be defined in corresponding actor role or not.
- An object model (O) is derived from an FM model of DEMO, which is shown in Fig. 6. There are two objects listed "Case" and "Proposal", with corresponding

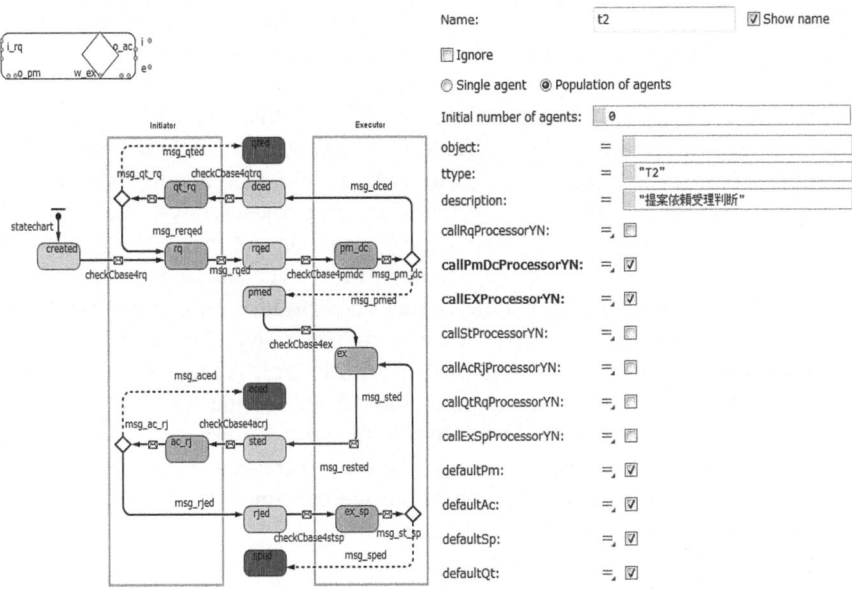

Fig. 5. T in DEMO++

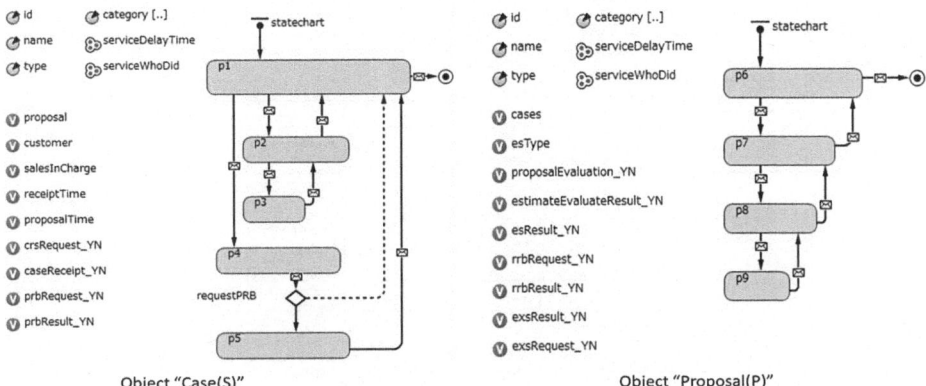

Fig. 6. Objects of "Company C" proposal and estimation process

properties. The properties "id", "name", "type", "category", "serviceDelay-Time", and "serviceWhoDid" are default properties for all of the objects. The remaining properties of "Case" are derived from DEMO FM "Case"; the properties of "Proposal" are derived from DEMO FM "Proposal". Object model described the possible states and state transition of the object.

- *Implementation part* (as shown in Fig. 4) defines the detailed execution of the onto-logical acts by considering the execution steps, the required resources, and the delays caused by execution, coordination and information transferring. The implementation part of DEMO++, including the resource (R) model and actor role component (AR component), need to be manually added.

- *Resource Model.* In this project, the actors are defined as active resource types who can play the actor roles. In Table 1. Actor-actor role-function mapping table (Case D) the following aspects are described: First, the table lists the eight types of actors from R1-R8, and nine types of actor roles A1-A9, according to the organizational structure of "Company C"; Second, all these actors are mapped into actor roles in DEMO. For example, actor "r1" (salesperson) takes on the actor role, meaning that this type of staff needs to take on these responsibilities; Third, the relationships between the actor roles and the functions that they support are defined. For example, the actor roles A1, A2 and A3 are utilized for the function F1 (case reception); A4 and A5 are for the function F2 (case risk evaluation); the actor roles A6 are for the function F3 (proposal and estimation); and A7–A9 are defined for the function F4 (proposal evaluation). Using an actor-actor role-function mapping table, we can understand the "who coordinates" and "coordinates for what" questions.

Table 1. Actor-actor role-function mapping table (Case D)

Actors								Actor Roles		Functions			
Sales Dep.			R&D Dep.	RB	QA Dep.	CEO	ES Office						
R1 Sales	R2 Sales TL	R3 Sales PM	R4 R&D department	R5 RB office	R6 Quality assurance department	R7 Executive officer	R8 Executive symposium office			F1	F2	F3	F4
T1.E													
T2.O													
T4.O,T4.R													
T5.O (info)		T5.O						A1	case responser				
T6.O													
T7.O													
	T2.E	T3.O						A2	case acceptance responser				
T3.O (info)													
	T3.E	T3.E	T3.E					A3	high-risk-case acceptance responser				
		T4.E						A4	proposal risk estimator				
	T5.E	T5.E	T5.E	T5.E (info)	T5.E			A5	high-risk-proposal risk estimator				
T6.E								A6	proposal completer				
	T7.E							A7	proposal evaluator				
T8.O (info)													
	T8.E	T8.E	T8.E	T8.E (info)	T8.E			A8	high-risk-proposal evaluator				
		T9.O (info)			T9.O								
						T9.E	T9.E (info)	A9	supper-high-risk-proposal evaluator				

- *Actor Role Components.* Implementation is related to the "how to coordinate" question. It is defined by the execution steps that are associated with the acts within the corresponding actor roles. To connect an ontological act with the implementation details, it is necessary to consider whether this act need to be expanded into a series of implementation details in its processor, the execution unit of an act. If so, the details of acts are defined according to real-world processes in the corresponding actor role. Actor role A1 is shown as an example in Fig. 7. A1 is the actor role of the "case responder" who is the executor of T1; they respond to [T1/pm], [T1/ex], [T1/dc], and [T1/st] and are the initiator of T2,

T4, T5, T6 and T7, which responds to rq and ac/rj for each transaction type. As described in Fig. 7, the implementation details for the acts, which must be executed in the processor, are described whereby the input defines the required act (the beginning of an act), and the output defines the status after proceeding with the act (the state after act). For example, act [T1/pm] is expanded in its processor A1 between input_pmdcT1 and output_pmdcedT1. The detailed execution process includes the business-level action "b_caseReceipt" and the decision-making act "CaseReceipt_YN", which determines whether to promise or decline a request according to the action rule defined in the AM.

- Both ontology elements and implementation elements are placed and connected in the *Main* part (as shown in Fig. 8). Transaction types, transaction types and actor roles, objects can communicate through the connections.

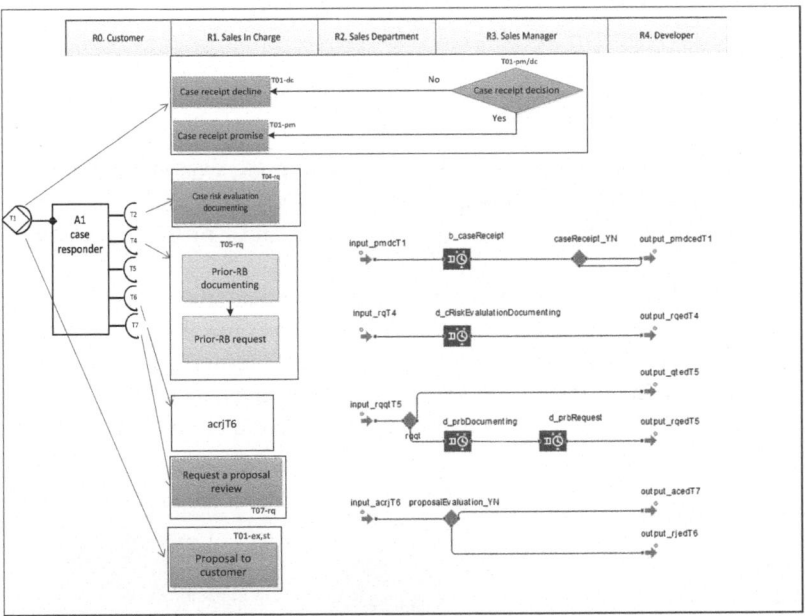

Fig. 7. Implementation details of the actor role "A1"

3.4   Simulation in AnyLogic Platform

The DEMO++ is designed as a conceptual model for simulation, that it can be translated into any simulation models. In this research AnyLogic is selected as the execution platform for its hybrid simulation capability. DEMO++ based AnyLogic simulation model is built on an "DEMO++ library for AnyLogic" that the author developed [4]. It completely follows the DEMO++ definition.

Simulation Results, Analysis, and Suggestions. The data was collected by (1) interviewing related staff, (2) checking documents and records and (3) assigning according to common characteristics.

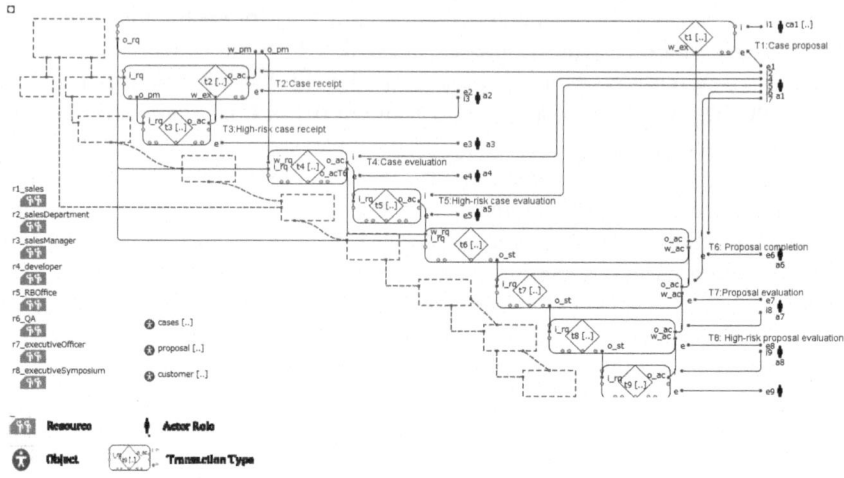

Fig. 8. Main of "Company C" proposal and estimation process

By simulation, we obtain average case reception times, whereby 40 % of the cases require less than three days; 45 % require less than seven days; and 13.9 % of the cases require more than eight days. For the proposal time, most of the cases require more than 10 days.

As shown in the top-left frame of Fig. 9. (1) Utilization of resources, the bottleneck is the sales department; approximately 96.6 % of salespeople are occupied most of the time. Therefore, cases and proposals must wait for this resource to be released. From (2) Total time spent in business level, information level and data level (B-I-D-level), we can note that only 39 % of a salesperson's time is used in B-level business; the other

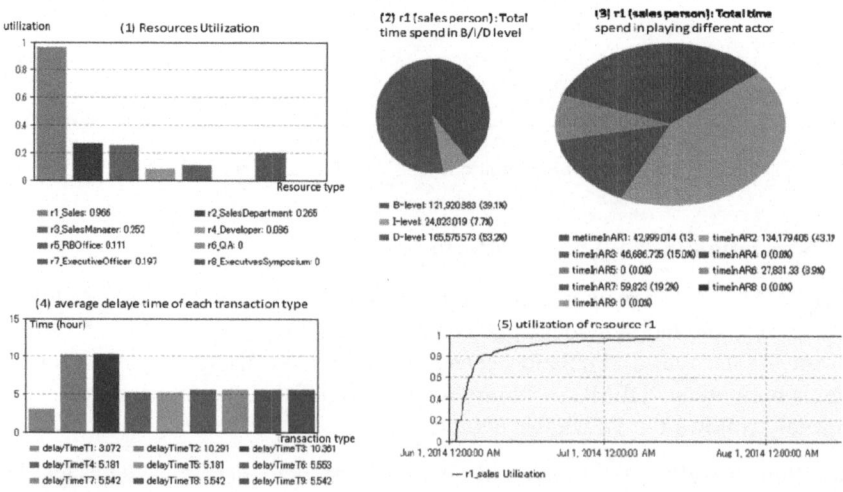

Fig. 9. Simulation results

61 % of a salesperson's time is used for information-level work and data transfer. According to "(3) Total time spent taking on different actor roles", the times r1 spent taking on different actor roles are as follows: 13 % in A1, 43 % in A2, 46.7 % in A3, 27.83 in A6 and 19.2 % in A7. As shown in "(4) delay time of each transaction", T2 and T3 are easy to delay because "r1" spends most of their time taking on these two actor roles.

To conclude, the problem of delay is mainly caused by the high utilization rate of a salesperson. However, the causes of the high utilization can be analyzed from two perspectives:

(1) Analyzing the Resources for Coordination:

- **(S1) Resource Allocation:** The most basic causes of the bottleneck problem may be a result of the resource allocation problem. Such problems can be solved through an optimization analysis. For example, in the case of "Company C", through an optimization analysis, the author observed that based on the original incoming rate and skill level, the enterprise does need not 25 salespeople; instead, the use of 40 salespeople would improve the proposal time and allow most cases to be completed within 10 days.
- **(S2) Effects of Resource Properties:** As an active resource, the personality, characteristics, skill level, knowledge structure, and social networks of an actor influence the effectiveness and efficiency of taking on an actor role. For example, in the case of "Company C", improving the skill levels of the salespeople reduces the time required for each case (related to type 2.4); the current percentage of A-level and C-level salespeople is 20 %. If some of the C-level salespeople can improve their skill level to the B-level and if some of the B-level salespeople can improve their skill level to the A-level, only 8 % of the salespeople will be C-level salespeople, with 34 % A-level sales people. Thus, these human resources reduce the r1 occupation time by 1 %–2 %.

(2) Analyzing the Coordination: From the simulation results, we can see that the coordination is not sufficiently efficient; therefore, the sales people are taking on too many information-level and data-level tasks. In addition, they are simultaneously taking on several actor roles.

- **(S3) Who Coordinates:** Add assistant staff to help perform the info-logical and data-logical tasks so that a time-constrained salesperson can concentrate on their value-added B-level work. The simulation suggests that the enterprise should assign another 10 assistant staff members to perform the info-logical and data-logical tasks, which will have the same effect on the result as solution S2.
- **(S4) How to Coordinate:** Change the process to reduce the time necessary for information transfer and documentation. Based on our simulation, for example, if the data-level work can be reduced by 20 %, most of the cases can be proposed within 10 days without adding resources or within only 5 days even if more salespeople are added.

Feedback. The solutions were proposed to "Company C". Their feedback regarding the results and the methodology is summarized as follows:

Concerning the Results:

1. This enterprise does not have sufficient skilled salespersons for the amount of work that must be completed. They are considering adding more workers, especially more skilled salespeople to remedy the situation. From this perspective, S1, S2 and S3 can help in the decision-making process. About solution S4, they consider the solution to be a good idea to increase efficiency; however, they still must investigate how to properly apply this solution.
2. One of the biggest concerns is related to the confidence of the data that they provided. It is very difficult for the company to determine the time requirements or measurements according to DEMO concepts. Some of the data are not empirical and are simply theoretical in nature.
3. Another concern is the B\I\D-level time measurements. The think it is an interesting point. But the feedback also indicates that the percentage should be holistically compared to determine if this distribution is normal. This is related to the Japanese working style—Japanese prefer to spend more time on documenting tasks compared to other countries.

Concerning the Method:

1. "Company C" shows great interest in this method. They agreed that it is different from traditional simulations, especially in the manner whereby it analyzes different levels of time requirements. The method appears to be more useful than other current methods, and has the potential to analyze more aspects of enterprises comparing to traditional simulations that can just measure.
2. However, more research and case studies are required to investigate how to implement the results. Meanwhile, there are still many aspects that must be improved if this method is to be commercialized, e.g., the development interface, the model transformation tool and animations.

Suggestions and Their Expectations:

1. It would be more helpful if the simulation could assist in investigating the relationships that exist among the key factors to determine the most criticized factors or to show more relationships between the key factors to provide additional guidance.
2. The animation must be improved to show the state of the bottleneck transaction.
3. The B-I-D-level analysis is very interesting; this research should continue in this direction to determine if there are further results that this method can provide.

4 Discussion and Future Research

In this project, the DEMO++ based simulation is evaluated to investigate how it can assist in business process change. By giving DEMO models, the proposal and estimation process of "Company C" is clarified. Comparing to traditional workflow perspectives, the company obtained a better understanding on the business process. Furthermore, based on DEMO model, DEMO++ models are derived by applying transformation tools. After that, the corresponding simulation is developed using DEMO++ libraries for AnyLogic, that it avoided the development from blank. This methodology combined high level analyzing with implementation level measurements and evaluation. It provided all the functions that traditional process based simulation can provide with additional advantages that they did not provide.

Moreover, a result of the project, several solutions are proposed according to the DEMO analyzing and DEMO++ based simulation results. These solutions give some interesting viewpoints for analyzing business process changes. More important, the types of solutions and the way of analyzing can be used as guidelines for the other DEMO++ based simulation projects as well.

The project also exposed limitations that need to be improved in the future researches:

- The implementation model. Currently, it need to be manually mapped into implementation models as execution steps and decision points. This process is still the most complex part in developing simulation. We need to provide a solution to make the mapping semi-automatically processed in the future research.
- The simulation and animation need to be improved. The current interface is designed for automatically model transformation. We can provide different animation interface according to different simulation objects. Moreover, current simulation is still weak in analyzing enterprise changes and provide solutions, for example the necessary changes that caused by organization structure changes. The further research need to consider more about how to make use of the current structure to provide more perspectives for analyzing changes.
- Only standard transaction pattern is focused. In certain real-world cases, it must simulate exceptions such as cancel or redo. The standard transaction pattern must then be expanded to describe the complete transaction pattern (with decline, reject and cancellation) to make the simulations more realistic and comprehensive.

References

1. Beer, M., Eisenbach, R.A., Spector, B.: Why change programs dont produce change. Harv. Bus. Rev. **68**(6), 158–166 (1990)
2. Kaufman, R.S.: Why operations improvement programs fail: four managerial contradictions. Sloan Manag. Rev. **34**(1), 83–93 (1992)
3. Dietz, J.L.G., Hoogervorst, J.A.P.: Enterprise Engineering (2014)
4. Liu, Y., Business process change analysis and business process simulation in the context of enterprise engineering (Dissertation), pp. 1–170. Tokyo Titech, Tokyo (2015)
5. Greasley, A., Barlow, S.: Using simulation modelling for BPR: resource allocation in a police custody process. Int. Oper. Prod. Manag. **18**(9/10), 978–988 (1998)
6. Du, X., Gu, C., Zhu, N.: A survey of business process simulation visualization. In: Proceedings of 2012 International Conference on Quality, Reliability, Risk, Maintenance, and Safety Engineering, pp. 43–48 (2012)

7. Jahangirian, M., Eldabi, T., Naseer, A., Stergioulas, L.K., Young, T.: Simulation in manufacturing and business: a review. Eur. J. Oper. Res. **203**(1), 1–13 (2010)
8. Barber, K.D., Dewhurst, F.W., Burns, R.L.D.H., Rogers, J.B.B.: Business process modelling and simulation for manufacturing management: a practical way forward. Bus. Process Manag. J. **9**(4), 527–542 (2003)
9. Reijers, H.A., Liman Mansar, S.: Best practices in business process redesign: an overview and qualitative evaluation of successful redesign heuristics. Int. J. Manag. **33**(58), 283–306 (2005)
10. Chwif, L., Banks, J., de Moura Filho, J.P., Santini, B.: Framework for specifying a discrete-event simulation conceptual model. J. Simul. **7**(1), 50–60 (2013)
11. Turnitsa, C., Padilla, J.J., Tolk, A.: Ontology for modeling and simulation. In: Proceedings of the 2010 Winter Simulation Conference, pp. 643–651 (2010)
12. Zeigler, B.P., Kim, T.G., Praehofer, H.: Theory of Modeling and Simulation, 2nd edn. Academic Press, San Diego (2000)
13. Dietz, J.L.G.: Enterprise Ontology-Theory and Methodology. Springer, Berlin (2006)
14. Barjis, J.: Automatic business process analysis and simulation based on DEMO. Enterp. Inf. Syst. **1**(4), 365–381 (2007)
15. Van Kerve, S.J.H.: Ontology driven enterprise information systems engineering. Ph.D. thesis (2012)
16. Liu, Y., Iijima, J.: Business process simulation in the context of enterprise engineering. J. Simul. **9**, 1–14 (2015)

On Complexity, Transformation
and Modeling

On the Role of Complexity for Guiding Enterprise Transformations

Jannis Beese[⊠], Stephan Aier, and Robert Winter

Institute of Information Management,
University of St. Gallen, St. Gallen, Switzerland
{jannis.beese, stephan.aier, robert.winter}@unisg.ch

Abstract. While there is a general agreement on the need for tools, which guide the evolution of complex organizational systems, and while there already exists a wealth of tools and approaches for the measurement and management of complexity, it seems that in practice these approaches often fail to achieve the desired impact during transformation processes. Based on focus group data and based on related literature, we analyze the factors that hinder current complexity management systems from guiding enterprise transformations and contribute a set of design principles, which address these factors. In particular, it is important to be aware of the context, to use a consistent ontology, to pay attention to visualization and to raise awareness and support.

Keywords: Complexity · Complexity management · Complexity measurement · Enterprise transformation

1 Introduction

Large enterprises need to continuously undergo transformation in order to adapt to varying external conditions. For large enterprises transformations comprise a series of local changes within the organization in order to cope with new and evolving requirements [1, 2]. While these local adaptations manage to temporarily fulfill the requirements, a series of such changes across different parts of the organization leads to unplanned and suboptimal states of the entire organization as a whole [3]. Inconsistencies, unnecessary redundancies or dependencies are introduced, which are typical drivers of complexity. This complexity in turn prevents people from recognizing a not only locally but globally optimal way to adapt and hinders efficient operation in the resulting state. Thus, there is a sustained practical [4–6] and academic [7–9] interest in complexity management and the development of underlying complexity measurement systems, which assist businesses in guiding transformations in a way that avoids unnecessary complexity.

There are multiple and diverse understandings of complexity [10–13] and the criteria for recognizing and measuring it vary both in terms of perspective and context [14–16]. Therefore, guidance on a higher level is required in order to coordinate developments within large enterprises with regard to complexity, so that inconsistencies and unnecessary redundancies are avoided or removed [17–20]. Resulting complexity

© Springer International Publishing Switzerland 2015
D. Aveiro et al. (Eds.): EEWC 2015, LNBIP 211, pp. 113–127, 2015.
DOI: 10.1007/978-3-319-19297-0_8

management methods rely on measurement systems that assess the current situation and evaluate the success of ongoing transformations [21].

However, it still seems that these complexity measurement systems often fail to achieve the desired impact during enterprise transformations: Various case studies show, that complexity management is considered important, yet not adequately supported [4, 6, 22]. For example, ATKearney report 84 % of companies recognizing complexity as a key cost driver, but these companies do not feel that they have "sufficient tools and systems to ensure continuous monitoring and controlling of complexity" [22].

This disparity between the effort spent on complexity measurement and management systems, and their perceived impact leads to the following research question:

Which principles should guide the design of complexity measurement systems in order to be useful for steering enterprise transformations?

We use a design science method following Peffers et al. [23] to tackle this question. Following Fischer [24], an abductive approach, based on focus group data, literature and conceptual analysis, is used to construct a set of design principles as target artifacts [25]. In order to identify and motivate our problem, we first identify seven factors that hinder complexity measurement systems from being used as guidance for enterprise transformations. We then propose four design principles for complexity measurement systems, which attempt to address these factors.

The rest of this paper is organized as follows: In Sect. 2 the conceptual foundations are laid out and an overview of related work is given, focusing on actual approaches to measuring and managing complexity. Building on these frameworks, in Sect. 3 a set of factors that inhibit their operationalization is presented and analyzed. This is used in Sect. 4 in order to derive a set of principles, which should guide the design and operationalization of complexity measurement and management systems in organizations. Section 5 discusses the scope and applicability of these principles, their limitations and potential implications, and gives an outlook on future work.

2 Conceptual Foundations and Target Artifact

In this paper we focus on the impact of complexity measurement on enterprise transformations. Measurement is not a goal in itself though, so in order to evaluate the effect it needs to be integrated in complexity management and organizational/structural decision processes that aim at sustaining a businesslxs ability to act efficiently, by transforming it in response to new requirements. This also makes sense from the other perspective: Complexity management methods rely on measurement systems in order to assess the current situation and in order to evaluate the success of current transformations [21].

In the context of this research, i.e. in dealing with large enterprises, complexity is in essence a human problem, meaning the inability of a person to make decisions and to take corresponding actions that guide the enterprise as a whole towards a globally optimal state due to too much or too complex information [15, 26]. While this interpretation differs distinctly from some descriptive and rather technical definitions (e.g. space or running-time complexity), it extends to the usual complexity measures for organizational and IS structures [16] in a natural way: An organizational entity, or a model thereof, with, for example, more components and relations among them, is often harder to analyze and understand, so that it will in turn be more difficult for a decision-maker to make the appropriate changes and to transform the overall system in an efficient and effective way. A typical example is the recognition and removal of unnecessary redundancies: It is hard from a local perspective to identify whether a given object is redundant or what the effects of its removal on the entire organization would be.

The reason for taking this perspective on complexity is that the target artifact of this research is not a precise definition of complexity, but an explanation of why the currently employed tools and methods often fail to achieve the desired impact during enterprise transformations and how this can be addressed.

Complexity measurement and management are generally employed to make businesses more agile, efficient, or robust, and to this end there exist very elaborate frameworks which support these goals in different environments or on different levels of the organizational hierarchy [21]. The design and implementation of these measurement and structural models are dependent on the context, defined by

(1) the objects to be evaluated (e.g. IS complexity, organizational complexity, task complexity or product complexity)
(2) the targeted users (e.g. IT managers, product designers, department heads, steering committees or enterprise architects)
(3) the goals of complexity management (e.g. agility, efficiency or robustness)
(4) environmental factors (e.g. industry factors, technological advances or market factors).

A good overview of common approaches to complexity measurement of different organizational entities is given in [16]. Even within a given category of objects, the differences in the operationalization of the measures are evident ([16], pp. 49–51). Additionally, a number of researchers are using ideas and analogies from cybernetics and complexity theory for analyzing complex systems in businesses, and although these approaches are not without criticism (cf. [27]), they introduce another set of very different ideas and approaches [8–10, 28, 29]. It is an important but difficult task to combine these approaches in a structured way to coherent ontological models [30].

Similar concepts exist for the management of complex projects, which aim at providing structured approaches for dealing with complexity or at reducing unnecessary complexity [18–20]. Recognizing and analyzing the given internal and environmental complexity is a prerequisite for the effective application of these techniques.

Different goals, design approaches, scopes and external factors lead to this diverse set of tools and methods. Therefore it does not seem appropriate to develop a "one-size-fits-all" solution for dealing with complexity [31]. Instead, multiple approaches can coexist within one organization, for example in order to deal with differences from

- necessary or external versus unnecessary or internal complexity
- "difficult", "complex" and "chaotic" systems [32]
- innovative and stable environments [33, 34].

These approaches, however, still need to be implemented, executed, supported and communicated in a coherent way in order to achieve maximum impact. We therefore propose a set of design principles, which serve as a guideline for these processes.

3   Analysis of the Problems of Complexity Measurement Tools

An analysis of the gap between the status quo of complexity measurement and management systems, and their perceived impact lends itself to a design and evaluation process involving focus groups: Experts from different organizations need to be involved in order to avoid the target artifact being only applicable in a specific case, but the nature of the research question calls for a deeper understanding and discussion [35, 36]. Hevner and Chatterjee ([35], pp. 123–124) point out, that a design science research approach involving focus groups

- allows "the researcher to clarify any questions about the design artifact as well as probing the respondents on certain key design issues",
- allows "deeper understandings, not only on the respondents' reaction and use of the artifact but also on other issues that may be present in a business environment that would impact the design" and
- allows "the emergence of ideas or opinions that are not usually uncovered in individual interviews",

all of which are important requirements for this research.

A series of three two-day workshops was held during June 2014 and February 2015 involving 16, 13 and 13 enterprise architects and high-level IT managers from ten different companies, respectively. The companies were mostly operating in banking and insurance, but also in logistics and utilities. According to the Global Brand Simplicity Index, the insurance industry is by far the most complex industry, with utilities and banking not too far behind ([37], p. 16). The size of the focus group allowed for an analysis of different ideas and viewpoints, while still being small enough for an in-depth discussion of more complicated questions. While the first two workshops were of an exploratory nature, with a focus on identifying, analyzing and grouping complexity factors, the final workshop had a confirmatory focus on evaluating potential solutions.

We identified a set of seven factors, which hinder the effective usage of complexity measurement and management systems. For every such factor we now explain the reasoning behind it before stating it together with a description, an example and any practical implications.

As complexity arises not as a local phenomenon that is easy to understand and explain, but instead results from the interactions that occur in larger systems, the attempts at making it tangible and manageable often exhibit the same complexity: The assessment process is difficult, involves diverse inputs and the resulting evaluations are hard to understand, interpret and communicate [14–16]. This in turn prevents people from using such tools during decision processes (Table 1).

Table 1. Factor: Complexity of the measurement system

1. Complexity of the measurement system	
Description	The measurement systems themselves are difficult to understand.
Example	Organizations provide various perspectives on complexity (e.g. IT, strategy, organizational) with different, complicated models that comprise aggregated and calculated measures, which are hard to explain and comprehend.
Implication	A large effort is required in order to understand and use the measurement system. The system is only accessible for a small group of users. Most people are not able to support their actions with it or even avoid using the system altogether.

There is only a very general agreement on the concept of complexity, which varies among people with different backgrounds [10–13]. When building measurement and management systems this lack of a common understanding leads to different interpretations of the same terminology. This in turn makes the design process more difficult and later on makes it hard to communicate justifications based on the resulting systems (Table 2).

Table 2. Factor: Unclear terminology

2. Unclear terminology	
Description	Terminology is used differently by different people or in different contexts.
Example	Often classification leads to problems: To which area or category does a given object or measure belong? Another example is that the precise understanding of typical goals of complexity management, such as agility or flexibility, often varies.
Implication	Decisions based on the obtained assessments are hard to communicate and its usefulness for the evaluation of potential actions will be limited.

Since complexity is an emergent property with interrelated causes, it is often hard or even impossible to find a definite indicator of complexity, which in all cases is apt for the given intent [10, 29]. Thus there exist very different measures, but their applicability depends both on the context and the specific goals [16]. In practice though, existing measures are often applied without considering their aptitude (Table 3).

Often some indicators for complexity are difficult to measure. The reasons vary and frequently include problems with data ownership, undocumented or outdated information, missing cooperation from stakeholders or general efforts and costs required (Table 4).

As mentioned in Sect. 2, differences in the observed systems and the resulting properties as well as differences in the pursued goals lead to very diverse approaches to measurement and management [14–16]. This is a problem if the resulting tools are structured and presented in an inconsistent fashion, thus making it hard to compare and act upon the obtained results (Table 5).

Table 3. Factor: Measures inapt for goals

3. Measures inapt for goals	
Description	The operationalized measures are not feasible or their relations to the targeted goals are unclear or not defined.
Example	Are lines of code an adequate measure for software complexity and what can be achieved based on that measure? Is size (e.g. amount of data) really relevant in all contexts, or only for certain organizations, services or products?
Implication	The resulting assessments do not match the purpose. People are not able to support decisions with assessments from the measurement system and show resistance against its usage and operation.

Table 4. Factor: Ability to obtain measures

4. Ability to obtain measures	
Description	Gathering certain measures requires inadequate effort or relies on missing coordination and cooperation.
Example	Is there up-to-date information available which will be maintained in the future? Often data is not voluntarily shared. How does one measure dependencies or connectedness? Is this data available?
Implication	The measurement systems cannot be developed and operated as originally planned and defined.

Table 5. Factor: Inconsistent presentation

5. Inconsistent presentation	
Description	Complexity assessments for different systems use a different presentation or different scales.
Example	Often several dashboards report complexity assessments for different areas, but they use different scales for the results, a different color coding or different layouts.
Implication	It is hard to communicate results and compare different options and systems.

Not being easily tangible and inherently hard to measure, systematic approaches for dealing with and evaluating complexity are often seen as unnecessary overhead or unwelcome monitoring. Furthermore, as these approaches span large parts of an organization, it is often unclear who is responsible for the continuous operation and development and for obtaining the required measures. This lack of support and responsibilities also leads to a very limited visibility of existing assessments and reports (Table 6).

An overall evaluation of the complexity of a system under scrutiny can be hard to explain, i.e. it is unclear which factors lead to the assessment and why one system is or is not considered complex. This makes it difficult to derive and evaluate key actions to be taken in order to improve system behavior [30] (Table 7).

Table 6. Factor: Lack of support and awareness

6. Lack of support and awareness	
Description	People are unwilling to support the operation and development of complexity measurement systems or are unaware of existing reports. They doubt the general benefit or are afraid of potential changes and implications.
Example	As complexity measurement systems initially introduce an additional effort for their construction and operation, people are unwilling to support these systems. Often there is a general resistance against a systematic collection of complexity indicators, which are used to monitor organizational or even individual performance.
Implication	Increased resistance against development and adaptation of the complexity measurement systems, along with little usage.

Table 7. Factor: Inexplicable results

7. Inexplicable results	
Description	Complexity assessments are not transparent and hard to explain.
Example	Complexity assessments are represented by a single number. It is unclear, what exactly this number means or what can be done in order to reduce or deal with this complexity.
Implication	It is not possible to derive actions, which might improve system behavior and efficiency.

The seven presented factors are not independent: While most of them positively reinforce each other (e.g. the complexity of the measurement system may result in inconsistent and unclear terminology, and vice versa, unclear terminology makes the measurement system appear more complex) or are unrelated, there are two factors, (1) and (3), which might lead to conflicts: Measurement systems are complex, partly because the underlying measures need to fit to very different objects. Thus a clear method for aggregating different measures into simple results is needed, which reduces the perceived complexity of the measurement system. This is reflected in principle (B), calling for a consistent ontology in the development process.

4 Principles for the Design of Complexity Measurement Systems

We now present a set of four principles for the design of complexity measurement systems, which address the factors presented in the previous section. For this, we follow the meta-model of Aier et al. [38]. Even though the context is different—Aier et al. describe a meta-model for principles for enterprise architecture development—the resulting artifact matches our requirements: We want to provide "the principles guiding [the] design and evolution" [39] of complexity measurement systems. Furthermore, the underlying theory is well developed (Fig. 1).

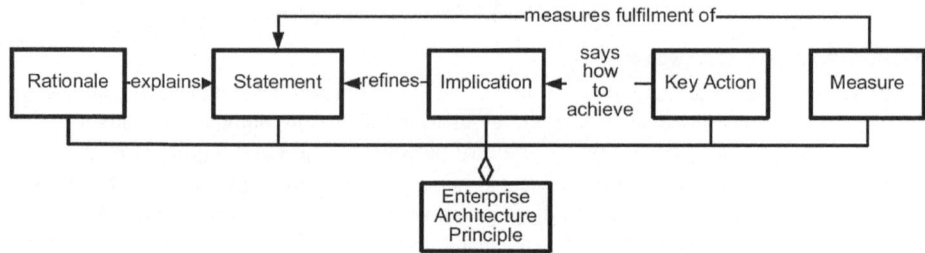

Fig. 1. Principle meta-model of Aier et al. (2011)

The principles are presented according to the following structure:

The statement (what is the goal?) itself is provided, along with the rationale (why should it be done?) behind it. Additionally, we describe measures (how the fulfillment of the principle is measured?) for the successful implementation of the principle and its implications (how can the goal be achieved?), i.e. general actions that follow from the principle. Company and scenario specific key actions (how can it be implemented in a specific case?) are not described, as these vary for each application of the principle [40].

The principles were developed by first analyzing potential solutions to the factors, i.e. what can be done or should have been done to avoid this, and then grouping and aggregating these solutions. This led to four core principles, which should guide the design and adaptation of complexity measurement and management systems:

A. Context-aware design process
B. Consistent ontology
C. Visualization
D. Awareness and support

This process allows for an easy mapping of the principles to the problem factors, which they attempt to address:

Following Venable et al. [41], we

- perform an early formative evaluation of these principles
- develop an artifact, which shall be useful for a heterogeneous group of stakeholders
- analyze socio-technical systems
- do not require strong rigor for our evaluation, as we do not develop a definite system or process, but rather provide guidance for its design [38, 39].

Thus, we select the recommended ex-ante, naturalistic approach to the evaluation of these principles by involving our focus group early during the design phase by paying attention to the applicability for real users and real systems [41]. Different approaches to the measurement of complexity within the participant's organizations were discussed with regard to partially solving the problem factors from the previous section. The proposed principles are therefore a result of an analysis of existing systems and an aggregation of proposed solutions to the problem factors.

Said and King ([42], Tables 8 and 9) identify the following factors as being most important for the usage of an IT system (such as the usage of a complexity measurement system supporting complexity management):

Table 8. Mapping the design principles to the addressed problem factors

					Factors			
		1	2	3	4	5	6	7
Principles	A			✓	✓		✓	
	B	✓	✓			✓		✓
	C	✓	✓			✓	✓	✓
	D		✓	✓	✓		✓	

Table 9. Principle A: Context-aware design process

A. Context-aware design process (3,4,6)	
Statement	The design and adaptation of complexity measurement systems needs to be aware of the specific context (goals, measured objects, target users).
Rationale	Depending on the context, certain types of complexity drivers are good or bad, or relevant or irrelevant and not every measure is applicable to every object and system. The design of the measurement system needs to adhere to these external limitations and requirements.
Implication	When designing a complexity measurement system, take the following into account: • The specific goals of complexity management for the evaluated objects • The targeted users • The type of objects under scrutiny • The available data
Measure	Indicators for the fulfillment of this principle give information about the fit of the measurement system on a local level: • Agreement of domain experts on the relevancy of the operationalized measures • Perceived relation between measures and goals • Average age of available data

Exogenous factors

(1) Compatibility: Fit of the system to the task performed.
(2) System rating: Perception of the overall characteristics of the system.
(3) Training: Extent of a user's knowledge and expertise with a system.

Endogenous factors

(4) Attitudes: How users feel towards the system.
(5) Relative advantage: Degree to which the system is more advantageous to other alternatives.
(6) Ease of use: Perceived usability of the system.

System rating is included in [42] as a general assessment of the quality of an IT-system and therefore not applicable in our case—a resulting principle would just be "build better systems". Relative advantage is also not applicable in the originally proposed sense: As there is usually only one complexity measurement system spanning the entire organization, no viable alternative systems exist. Principle (D) addresses this factor in some sense by explaining the benefits of using a complexity measurement system versus simply relying on intuition. The remaining factors map to the proposed principles: Compatibility is addressed by using a context-aware design process (Principle A) and by using a consistent ontology (Principle B) that explains the relation between the measures and the goals. Attitude is addressed by principle (D). Ease of use is addressed by the visualization principle (C), requiring a simple and easy to understand presentation. We are therefore confident that the proposed principles cover the most important approaches.

The first principle (A) addresses the fact that the targeted entities in an organization differ widely and, related to this, that the target states vary [43]. One needs to be aware of these differences during the design process in order to choose the right approach and tool for the given scenario and intent. This in turn increases the effectiveness of the resulting complexity management system, by ensuring that the measures are related to the goals (3) and are applicable in the specific context (4). It further makes it easier to explain why the measures are relevant (6), as they were chosen with the specific context in mind.

Principle (B) is concerned with the establishment of a consistent ontology, explaining and naming all involved objects, properties and their relations. Complexity in businesses is an issue, which is inherently hard to define and different people with different backgrounds will have a different understanding of what this means in detail. Furthermore, the relation between the targeted goals and the supporting systems is often unclear, inhibiting their usage during transformation: Pombinho criticizes, that "most methods used up to now to manage this complexity are not based on a transversal, coherent and concise conceptual model" [30]. Explaining the relevant objects and relations helps to resolve misunderstandings due to terminology (2) and thus make the resulting complexity management system easier to understand (1). Additionally, knowledge about the relations within the measurement system helps to interpret and explain complexity assessments on a more detailed level (7) and assist in recognizing and resolving inconsistencies (5), both of a technical and terminological nature (Table 10).

Principle (C) states that important results should be visualized in an aggregated, consistent and easy to understand way. While it should still be possible for experts to analyze the details of this aggregate result, a simple, graphical representation, which highlights important information, is essential. This not only hides the inherent complexity of the measurement system itself (1) and thus makes it more accessible for users (6), it also makes it easier to follow a consistent terminology (2) and presentation (5) and to explain and compare results (7) (Table 11).

Finally, Principle (D) addresses common issues with acceptance and usage, which result from people not being aware of existing systems and methodological capabilities or from a lack of clear responsibilities for driving the development and usage of complexity measurement systems. Thus, Principle (D) requires a clear plan the

Table 10. Principle B: Consistent ontology

B. Consistent ontology (1,2,5,7)	
Statement	Complexity measures should be based on a consistent ontology, which names and describes all relevant objects, properties and their relations.
Rationale	This principles stems from two main difficulties (see Sect. 3): 1. Complexity is inherently hard to define precisely. 2. The relations that lead to a complexity assessment need to be explained, in order to allow for actionable advice. These difficulties need to be overcome in order to allow people to work effectively with the complexity assessments and integrate them into transformation processes.
Implication	• Give a clear definition of relevant objects. • Describe relations between objects/measures. • Describe how the measures are related to the goals and to overall complexity assessments. • Identify and resolve potential conflicts. • Identify which measures are relevant in a given context.
Measure	• Defined terms • Are aggregated and calculated measures explained? • Are the measures mapped to the goals?

Table 11. Principle C: Visualization

C. Visualization (1,2,5,6,7)	
Statement	Results and explanations of complexity assessments should be presented in a simple, unified and consistent fashion.
Rationale	Complexity measurement systems must be accessible for users from different roles, organizations and with different knowledge. As such an easy to understand presentation is important. In order to ease discussion and comparison, this presentation of results should be consistent across all aggregation levels and contexts.
Implication	• Comply with corporate design. • Use the same format for all reports. • Present aggregated and detailed results similarly. • Highlight important results in an easy graphical way.
Measure	• Number of different templates/designs • Percentage of reports, which follow corporate standards • Perceived ease of use • Percentage of reports including simple, aggregated assessments

definition of responsibilities and for raising awareness, both for the necessity and benefits of the measurement and for potential applications of resulting assessments. This insures that irrelevant or misleading measures are detected early and can be corrected (3) along with terminological issues (2). Furthermore, it will be easier to gather support and cooperation (6), and makes it easier to obtain required measures (4) (Table 12).

Table 12. Principle D: Awareness and support

D. Awareness and support (2,3,4,6)	
Statement	Design of measurement systems needs to be supported by raising awareness and selecting people, who are responsible for driving usage and development.
Rationale	In order to have an impact, people need to be aware of the existence and potential applications of the measurement system. Additionally, the effort involved in the gathering measures needs to be justified by explaining the resulting benefits.
Implication	• Define clear responsibilities. • Explain the benefits and potential use cases. • Train people in the usage of the measurement system.
Measure	• Percentage of people aware of the system • Percentage of components with clear ownership • Access and usage statistics

5 Discussion and Outlook

As the development of the principles stems from real-world problems of practitioners, we are confident that these provide a useful guidance for the design of complexity measurement and management systems. The underlying problems do not come from a lack of interest or resources – all companies involved in the focus group employ sophisticated complexity management and reduction programs. The issue lies with actually generating an impact from there: The effect of these systems relies on them being used and supported by people in an organization.

Thus the underlying questions are: How can an understanding of complexity be introduced into a company, so that complexity assessments are used both intuitively and systematically support transformations? How can we guide this series of small, local changes so that it converges to an efficient global state of the business [44]. The proposed principles provide a first point for further discussion in this direction, by giving guidance on the development and adaptation of complexity management systems. Additionally, there are approaches to develop complexity management techniques based on insights from complexity theory, which not necessarily try to reduce complexity, but to manage it in an adequate fashion corresponding to the underlying system complexity [28]. The general problem though is likely not solved by the design and usage of a complexity management system alone, but also has strategic roots [45, 46] as well as connections to corporate culture and leadership [47, 48]. Additionally, while still requiring adequate support from complexity measurement systems, applying insights from complexity theory to management methods also might help to solve the problem of dealing with complexity [28]. This is outside of the scope of this paper, but presents an interesting area for future, related research.

The presented principles also would benefit from further, more detailed practical evaluation: As the focus groups consisted of enterprise architects and high-level IT-managers of large companies in different industries, which need to report and justify their investments in complexity management and supporting systems, we believe that the issues of general complexity measurement systems are addressed quite well. Nevertheless, complexity in the context of, for example, company strategy or product design is quite different to IS complexity or organizational complexity and it needs to be analyzed to which extend the principles apply to the former areas [11, 16].

References

1. Miller, D.: Environmental fit versus internal fit. Organ. Sci. **3**, 159–178 (1992)
2. Vessey, I., Ward, K.: The dynamics of sustainable IS alignment: the case for IS adaptivity. J. Assoc. Inf. Syst. **14**, 283–311 (2013)
3. Dooley, K.J., Van de Ven, A.H.: Explaining complex organizational dynamics. Organ. Sci. **10**, 358–372 (1999)
4. Heywood, S., Spungin, J., Turnbull, D.: Cracking the complexity code. McKinsey Q. **2**, 85–95 (2007)
5. Shane, J.S., Strong, K.C., Gransberg, D.D.: Project Management Strategies for Complex Projects. (2014)
6. Tanaka, H.: Toward project and program management paradigm in the space of complexity: a case study of mega and complex oil and gas development and infrastructure projects. Procedia - Soc. Behav. Sci. **119**, 65–74 (2014)
7. Nan, N.: Capturing bottom-up information technology use processes: a complex adaptive systems model. MIS Q. **35**(2), 505–532 (2011)
8. Gharajedaghi, J.: Systems Thinking: Managing Chaos and Complexity: A Platform for Designing Business Architecture. Elsevier, Amsterdam (2011)
9. Pellissier, R.: A proposed frame of reference for complexity management as opposed to the established linear management strategies. Int. J. Organ. Innov. **5**, 6–67 (2012)
10. Cooke-Davies, T., Cicmil, S., Crawford, L., Richardson, K.: We're not in kansas anymore, toto: mapping the strange landscape of complexity theory, and its relationship to project management. Proj. Manag. J. **38**, 50–61 (2007)
11. Dewar, R., Hage, J.: Size, technology, complexity, and structural differentiation: toward a theoretical synthesis. Adm. Sci. Q. **23**, 111–136 (1978)
12. Edmonds, B.: What is complexity? - the philosophy of complexity per se with application to some examples in evolution. In: Heylighen, F., Aerts, D. (eds.) The Evolution of Complexity. Kluwer, Dordrecht (1995)
13. Shalizi, P.C.R.: Methods and techniques of complex systems science: an overview. In: Deisboeck, T.S., Kresh, J.Y. (eds.) Complex Systems Science in Biomedicine, pp. 33–114. Springer, Heidelberg (2006)
14. Bosch-Rekveldt, M., Jongkind, Y., Mooi, H., Bakker, H., Verbraeck, A.: Grasping project complexity in large engineering projects: the TOE (technical, organizational and environmental) framework. Int. J. Proj. Manage **29**, 728–739 (2011)
15. Geraldi, J., Maylor, H., Williams, T.: Now, let's make it really complex (complicated) a systematic review of the complexities of projects. Int. J. Oper. Prod. Manage. **31**, 966–990 (2011)
16. Weidong, L.: Complexity of information systems development projects: conceptualization and measurement development. J. Manage. Inf. Syst. **22**, 45–83 (2005)
17. Ashmos, D.P., Duchon, D., McDaniel, Jr., R.R.: Organizational responses to complexity: the effect on organizational performance. J. OrgChange Mgmt **13**, 577–595 (2000)
18. Remington, K., Pollack, J.: Tools for Complex Projects. Gower Publishing Ltd., Hampshire (2007)

19. Whitty, S.J., Maylor, H.: And then came complex project management. Presented at the 21st IPMA World Congress on Project Management (2007)
20. Whitty, S.J., Maylor, H.: And then came Complex Project Management (revised). Int. J. Proj. Manage **27**, 304–310 (2009)
21. Geraldi, J.G.: What complexity assessments can tell us about projects: dialogue between conception and perception. Technol. Anal. Strateg. Manag. **21**, 665–678 (2009)
22. Study on: Complexity Management – Chances amid the crisis (2009) http://www.mycomplexity.com/complexity_management_publications/Complexity_Management_Study_Results_sent_internally.pdf
23. Peffers, K., Tuunanen, T., Rothenberger, M., Chatterjee, S.: A design science research methodology for information systems research. J. Manage. Inf. Syst. **24**, 45–77 (2007)
24. Fischer, C., Gregor, S., Aier, S.: Forms of discovery for design knowledge. In: ECIS 2012 Proceedings (2012)
25. Gregor, S., Hevner, A.R.: Positioning and presenting design science research for maximum impact. MIS Q. **37**, 337–356 (2013)
26. Schlindwein, S.L., Ison, R.: Human knowing and perceived complexity: implications for systems practice. Emergence: Complex. Organ. **6**, 27–32 (2004)
27. Ison, R., Schlindwein, S.L.: History repeats itself: current traps in complexity practice from a systems perspective. Presented at the 12th Australia New Zealand Systems Society (2006)
28. Benbya, H., McKelvey, B.: Toward a complexity theory of information systems development. Info Technol People **19**, 12–34 (2006)
29. Berry, B.J.L., Kiel, L.D., Elliott, E.: Adaptive agents, intelligence, and emergent human organization: capturing complexity through agent-based modeling. PNAS **99**, 7187–7188 (2002)
30. Pombinho, J., Aveiro, D., Tribolet, J.: Value-oriented specification of service systems: modeling the contribution perspective of enterprise networks. Int. J. Inf. Syst. Serv. Sect. **7**, 60 (2015)
31. Winter, R.: Construction of situational information systems management methods. Int. J. Inf. Syst. Model. Des. **3**, 67–85 (2012)
32. Snowden, D.J., Boone, M.E.: A leader's framework for decision making. Harv. Bus. Rev. **85** (11), 68 (2007)
33. Brown, S.L., Eisenhardt, K.M.: Competing on the Edge: Strategy as Structured Chaos. Harvard Business Press, Boston (1998)
34. Yayavaram, S., Chen, W.-R.: Changes in firm knowledge couplings and firm innovation performance: the moderating role of technological complexity. Strat. Mgmt. J. **36**, 377–396 (2015)
35. Hevner, A., Chatterjee, S.: Design science research in information systems. In: Hevner, A., Chatterjee, S. (eds.) Design Research in Information Systems, pp. 9–22. Springer, US (2010)
36. Tremblay, M.C., Hevner, A.R., Berndt, D.J.: The use of focus groups in design science research. In: Hevner, A.R., Chatterjee, S. (eds.) Design Research in Information Systems, pp. 121–143. Springer, US (2010)
37. Siegel + Gale: Global Brand Simplicity Index 2014. Siegel + Gale (2014)
38. Aier, S., Fischer, C., Winter, R.: Construction and evaluation of a meta-model for enterprise architecture design principles. In: Wirtschaftsinformatik, p. 51 (2011)
39. TOGAF Version 9.1. The Open Group (2011)
40. Hoogervorst, J.: Enterprise architecture: enabling integration, agility and change. Int. J. Coop. Info. Syst. **13**, 213–233 (2004)
41. Venable, J., Pries-Heje, J., Baskerville, R.: A comprehensive framework for evaluation in design science research. In: Peffers, K., Rothenberger, M., Kuechler, B. (eds.) DESRIST 2012. LNCS, vol. 7286, pp. 423–438. Springer, Heidelberg (2012)

42. Al-Gahtani, S.S., King, M.: Attitudes, satisfaction and usage: Factors contributing to each in the acceptance of information technology. Behav. Inf. Technol. **18**, 277–297 (1999)
43. Harrison, D.A., Klein, K.J.: What's the difference? diversity constructs as separation, variety, or disparity in organizations. Acad. Manag. Rev. **32**, 1199–1228 (2007)
44. Levinthal, D.A., Warglien, M.: Landscape design: designing for local action in complex worlds. Organ. Sci. **10**, 342–357 (1999)
45. Hoogervorst, J., van der Flier, H., Koopman, P.: Implicit communication in organisations: The impact of culture, structure and management practices on employee behaviour. J. Manag. Psychol. **19**, 288–311 (2004)
46. Kurtz, C.F., Snowden, D.J.: The new dynamics of strategy: Sense-making in a complex and complicated world. IBM Syst. J. **42**, 462–483 (2003)
47. Lichtenstein, B., Uhl-Bien, M., Marion, R., Seers, A., Orton, J., Schreiber, C.: Complexity leadership theory: an interactive perspective on leading in complex adaptive systems. Management Department Faculty Publications. (2006)
48. Uhl-Bien, M., Marion, R., McKelvey, B.: Complexity Leadership Theory: Shifting leadership from the industrial age to the knowledge era. Leadersh. Q. **18**, 298–318 (2007)

On the Explanatory Capabilities of Enterprise Modeling Approaches

Monika Kaczmarek$^{(\boxtimes)}$, Alexander Bock, and Michael Heß

Chair of Information Systems and Enterprise Modeling,
Faculty of Business Administration and Economics, Institute for Computer Science
and Business Information Systems (ICB), University of Duisburg-Essen,
Universitätsstraße 9, 45141 Essen, Germany
{monika.kaczmarek,alexander.bock,m.hess}@uni-due.de

Abstract. The capability of an enterprise modeling approach to support the provision of knowledge on selected aspects of an enterprise may be apprehended as its *explanatory capability*. We argue that this capability encompasses two aspects: the capability to represent "the things happening in an enterprise" and the 'self-explanatory' capability that relates to the understandability of the approach and the resulting models. In this paper, we propose an analysis framework that can be used to assess the explanatory capabilities of enterprise modeling approaches. The framework is structured according to the four explanatory causes of Aristotle. We demonstrate the applicability of the framework by analyzing three selected enterprise modeling approaches.

Keywords: Enterprise model · Explanatory capability · Analysis framework

1 Introduction

Enterprise modeling (EM) builds on conceptual modeling to support the description, reflection upon, and (re-)design of various aspects of enterprises [1]. A number of EM approaches exist that offer different sets of modeling concepts from which a particular enterprise model can be created (see, e.g., [2]). It can be argued that a main role of an enterprise model is the provision of knowledge on selected aspects within, or related to, an enterprise. The capability of an EM approach to support knowledge expression of this sort may be apprehended as its *explanatory capability* [3]. The explanatory capability of an EM approach can be seen to involve two dimensions. The first one can be understood as the capability to describe "the things happening in an enterprise". This capability depends on the scope of the modeling concepts provided by the approach, which represent a means for reflecting phenomena of the considered domain in the service of specific goals. The second one may be termed the 'self-explanatory' capability, and relates to the understandability of the approach and the resulting models as perceived by its users. This capability depends on the characteristics of both the modeling approach and the involved actors.

© Springer International Publishing Switzerland 2015
D. Aveiro et al. (Eds.): EEWC 2015, LNBIP 211, pp. 128–143, 2015.
DOI: 10.1007/978-3-319-19297-0_9

The explanatory capabilities of EM approaches may be regarded as a key success factor for the usage of the resulting enterprise models. But although the importance of features linked to the explanatory capabilities of approaches are commonly acknowledged (see, e.g., [4–7]), studies explicitly investigating existing approaches in view of constructs such as 'explanatory capability' can hardly be found. Therefore, to contribute to the consolidation and evolution in the field of EM, in this paper, we continue earlier research [2] and analyze selected EM approaches by focusing on their explanatory capabilities. To this aim, we design an analysis framework and demonstrate its applicability by analyzing selected approaches. According to the classification proposed by [8, pp. 251–260], our analysis exhibits characteristics of a 'theoretical and conceptual investigation' supported by the 'feature comparison'. In turn, following the classification scheme proposed by [9, p. 98], the framework and the conducted analysis can be classified as 'vertically dominant' [9, p. 111], as we consider a small number of coarse criteria and apply them to compare a small subset of existing methods.

The paper is structured as follows. First, we discuss the notions of model and enterprise model, and position our work towards the work of others (Sect. 2). Next, the analysis framework is explained (Sect. 3) and applied to characterize selected approaches (Sect. 4). The paper concludes with final remarks.

2   Enterprise Modeling and Models

Models, often regarded as abstractions created with a certain purpose in mind [10], are perceived to be central to all forms of understanding (e.g., [10–12]). As reality is highly complex, models result from attempts to separate and isolate different phenomena and to identify key relationships among them [10,12]. In this context, a distinction has been suggested between 'ontological' and 'descriptive' complexity [13]. The 'ontological' complexity refers to the 'actual' complexity of phenomena (or a domain), whereas the 'descriptive' complexity focuses on the complexity of the description provided by the model [12,13]. If we consider a model as a 'construction' resulting from "purposeful abstraction of a domain" [14, p. 4], then a model allows to deal with the 'ontological' complexity of the domain by either fading out aspects not being relevant for a given purpose, or changing and adding some features to better achieve the particular purpose. On the other hand, a model itself is an artifact, i.e., a representation resulting from a purposeful construction [14, p. 4]. As such, it possesses its own level of 'descriptive' complexity, which, in order to foster understanding and communication, should be adjusted to the needs and abilities of prospective users.

Enterprise modeling "is the process of understanding a complex social organization by constructing models" [15, p. 18]. More precisely, an enterprise model is a *conceptual model* (e.g., [1, pp. 942–943]), hence a deliberate linguistic construction [14, p. 24]. Enterprise models, whether used descriptively or prescriptively [14], usually integrate conceptual models of information systems (IS) with conceptual models of organizational action systems [14]. Enterprise models are created using modeling languages, defined through their abstract syntax, semantics and concrete syntax [14]. Modeling languages might be regarded as a common language supporting communication between stakeholders [11, p. 28].

From the mentioned properties of models in general, two desirable capabilities of enterprise modeling approaches in particular emerge. First, EM approaches should enable to reduce the 'ontological' complexity of the phenomenon in question, hence an enterprise. Considering their defined purposes, they should offer the means (i.e., modeling concepts) for reflecting the relevant aspects of an enterprise. We consider this the first explanatory capability (i.e., the ability to explain 'an enterprise'). Second, still in view of the defined purposes, enterprise models should be understandable to prospective users. This means that the inherent 'descriptive' complexity of a modeling language, encompassing its modeling concepts (abstract syntax) and its representation (concrete syntax), as well as the general suggested way of thinking about the domain in question, should be adequately adjusted to the defined goals and target user groups. This capability is regarded as the 'self-explanatory capability'.

Considering the plethora of existing EM approaches that are available today, at least two questions follow: (1) *What are the constituents of the explanatory capabilities of EM approaches?* and (2) *Do the existing EM approaches significantly differ with respect to their explanatory capabilities?* A number of comparative analyses of enterprise modeling (EM), enterprise engineering (EE), and enterprise architecture (EA) approaches have been undertaken by various authors (e.g., [8, 16–19]). However, to our best knowledge, only Kirikova [3] undertook an attempt to analyze modeling approaches taking into account the concept of explanatory capabilities. As explained subsequently, we use this work as a starting point, for which we propose numerous augmentations and modifications. We also extend the analysis by considering research conducted in the fields of complexity science (e.g., [13, 20–22]), organizational theories (e.g., [23–26]), and existing works on the economics of modeling (e.g., [7]), understandability of models in general, and process models in particular (e.g., [4, 6, 27, 28]).

3   Analysis Framework

To answer the above-stated questions, we design and apply an analysis framework to systematize the general constituents of explanatory capabilities of EM approaches (Question 1), and guide the comparative analysis of concrete approaches building on these criteria (Question 2). Designing such a framework is a challenging task. Any framework runs the risk of imposing the designer's categories and perspectives on the analysis. With respect to the first explanatory capability, the framework needs to be general and abstract, but still meaningful enough, to allow for considering different views on an enterprise that can be followed by different modeling approaches, and without favoring any of them. With respect to the self-explanatory capability, the framework should account both for an artifact (an enterprise model) as well as the act of its creation and involved actors, and allow to tightly couple them with the targeted purpose.

Taking into account the goal of our research, the aforementioned concerns, and existing work (e.g., [3, 20–22, 29]), we utilize the scheme of the four causes of

Table 1. The causes and principles by Aristotle based on [30]

Cause	Explanation	Questions addressed and examples
Material	The cause "out of which the thing comes to be" [30]	What does a thing actually consist of? *Example:* The bronze a statue is made of
Formal	The definition of the essence; the form of the artefact to be	How are (prospective) constituents to be connected, shaped, or changed to form the thing? *Example:* The prospective shape of a statue
Efficient	The primary source of the change that achieves a thing, including human knowledge	How does change unfold? Who, and with which knowledge, is performing it? *Example:* The art, and the actual work of casting the statue
Final	The ultimate goal of a thing	What is the main idea standing behind a thing? *Example:* The purpose of the statue

Aristotle to structure the proposed framework. The scheme of Aristotle is highly general, consisting of four 'causes', which are thought to represent different facets of "an explanation for how something came to be" [30]. The causes are explained in Table 1. They are used as general lenses to examine explanatory capabilities of EM approaches, with the aim of identifying more specific constituents. We apply Aristotle's scheme as it offers a basis for justifying the framework elements in a manner which is general, comprehensible and plausible, at least to the degree that the philosophical scheme is. It allows us to abstract from the particular understanding of an enterprise and an enterprise model assumed by the various approaches, serving as a broad, overarching structure for arranging aspects from different theoretical lines of thought. This can reduce the potential of an arbitrary 'ad hoc' framework definition (cf. [29, p. 75]). However, although we apply the causes to identify elements of an analysis framework, we essentially develop our own interpretations, whilst taking into account existing work in various research fields mentioned below. Thus, we use the Aristotelian doctrine as a source of inspiration only, being aware of its ambiguities and that its relevance to science is still an unresolved and disputable issue (cf. [21]).

3.1 The 'causes' of an Enterprise

The Aristotelian causes were already used to analyze enterprises in the field of EM [3], IS and organization theories [29] and complexity science [20–22]. While sensible in their own right, these works involve interpretations and goals that are different from ours. We reassess these suggestions, and propose augmentations and modifications, as summarized in Table 2, and discussed below.

The material cause is to capture "that out of which a thing comes to be" [29, p. 71]. The question of what constitutes an 'organization' or an 'enterprise' is not a trivial one, and a subject of long-standing and ongoing debates in organization and management studies (cf. [23, pp. 15–18]). Already traditional views emphasize that an enterprise is not a natural phenomenon but exists only insofar as it is produced through collective efforts of individuals [31, p. 26]. At a basic level, it is understood either in an instrumental (describing the way collaborative work is divided and governed) or an institutional sense (describing an

Table 2. Results of applying Aristotle's 'causes' to the subject of an enterprise

Cause	Aspects of an enterprise	Comments
Material	Tangible and intangible resources, personnel and their skills, IT infrastructure, human communication and interaction	Elements of the action system and IS related to enterprise infrastructure; actions and language used
Formal	Organizational structures, hierarchies, business processes and rules	The way collaborative work is formally *and* informally divided and governed
Efficient	Activities performed by enterprise members, interaction and communication	Actual work done; means to track and measure it (e.g., metrics or indicators)
Final	(Declared) visions and missions, organizational goals, individual goals	Desired ends, whether officially stated, implicitly pursued, or individual ones

organization as a social system) (e.g., [32, p. 15]). The former view comprises a more or less explicit set of rules and norms; the latter embraces an entire socio-technical entity [32, p. 15]. When focusing on the latter view, it seems intuitive to assume that an 'enterprise' consists of individuals, material objects (e.g., buildings, products), and immaterial resources (e.g., services). Traditional views indeed regard these as at least one basic component of an enterprise [31, p. 29]. In line with this institutional view, Dalhbohm and Mandal [29, p. 71] essentially see the 'material cause' to refer to different kinds of resources (they use the term 'infrastructure'), including capital, technology, personnel, buildings, and indirectly systems of transport and finances. Similarly, Kaminska et al. [22, p. 13] understand the material cause to relate to resources, capabilities, and competences that can be used to perform various activities. While this interpretation of the Aristotelian cause is comprehensible, it is misleading to equate it with the "material from which an organization is made", as is stated by [29, p. 71]. When considering the instrumental organization of an enterprise, and the human activities in line with these coordinating measures, it can be argued that an enterprise as such "has no physical being", and exists solely as an "artifact of human cognition" [26, pp. 103–104]. As a result, it has been suggested that an "organization is a *product* of communication, and totally dependent on symbolic sense making through interaction for its mere existence" [26, p. ix]. Although this view has gained momentum rather recently [33], related propositions have been made in early work as well (e.g., [24, pp. 164–165]). To be able to account for the diversity of stances followed by the existing EM approaches, we do not adopt a particular conception of an 'enterprise' here. We acknowledge that enterprises are understood, inter alia, as clusters of material and immaterial resources, as well as matters made solely of human communication and interaction.

The formal cause describes that what defines the form of the considered phenomenon in question. This can be related to the way collaborative work is coordinated (i.e., the instrumental organization). The same interpretation is made by Kaminska et al. [22, p. 13] and Dalhbohm and Mandal [29, p. 71], who even regard the formal cause as the "organization as such". From a formal point of view, this relates to the organization's structure, system, and processes; i.e., to explicit rules defining the division and coordination of labor. The focus of early organization research has been placed on investigations and guidelines related to this formal dimension of organizing, including the analysis of determinants such as unity of command, specialization, span of control, and task definitions (e.g., [24, pp. 12–33]). In addition to the formal organization, however, it has been suggested that the informal organization (covering, e.g., informal roles and workarounds) shapes an equally relevant portion of how activities in an enterprise are performed (e.g., [32, pp. 13–14]). In line with [29, p. 71], we take the 'formal' cause to encompass both the formal and informal way of dividing work.

The efficient cause is mapped by [29, p. 71] to the daily activities performed by enterprise members, including process execution and resource usage, i.e., the actual "getting work done". More generally, it might be stated that the efficient cause comprises any kind of human activity, interaction, and communication that effectively brings forth an organization from day to day. In addition, Kaminska et al. [22, p. 13] note that the efficient cause might be captured by means of measures, indicators, or other tools, which track, e.g., used resources. While these tools are not themselves the efficient cause, they may be regarded as an (indirect) reflection of it. Thus, we argue that they should be considered as well.

The final cause is easily misleading when applied to an enterprise in a direct fashion. Kaminska et al. [22, p. 13] relate the final cause to strategic objectives, which would guide activities at all levels. It is added by [29, pp. 71–72] that it was "very unusual that organizations have a clear conception of their goals." Indeed, as has been emphasized in organization studies, goals cannot be simplistically attributed to an organizational entity (e.g., [25, pp. 26–43]). Instead it needs to be considered how goals emerge in social systems. This means that there might be officially declared visions, missions, and goals; as well as local, in part implicit, and personal goals [25]. All of these can be assumed to emerge in an interrelated manner, whilst influencing organizational behavior [24, 25].

3.2 Self-explanatory Capabilities

The self-explanatory dimension focuses on the extent to which the use of an EM approach supports communication and 'sense-making' processes by enabling to describe an organization in a manner that is intendedly understandable to different stakeholders. We argue that in order to analyze the self-explanatory capabilities, the procedure(s) applied to create an enterprise model, the modeling language, and the involved actors need to be considered (see Table 3).

A model is a purposeful abstraction of a domain. As has been indicated, organizational action systems are largely constituted by linguistic representation and communication processes. Creating an enterprise model consequently requires analyzing linguistic accounts of the domain of discourse [14, pp. 23–24]. Therefore, following [3], *the material cause* is seen to refer to the way an EM approach suggests to analyze a domain of discourse in order to acquire the information necessary for creating an enterprise model. The way of obtaining linguistic accounts of a domain can vary. For example, a modeler might obtain information by interviewing stakeholders, or by accessing available documents or information systems [6]. Alternatively, the modeling approach could also empower domain users [5] and allow them to become active modelers.

The formal cause concerns the modeling language and can be related to at least three aspects. Firstly, it can be related to the correspondence between the semantics of provided modeling concepts and (natural) language concepts from the domain of discourse with which intended users might be familiar [3], [14, p. 25]. In other words, this relates to the 'closeness' of a modeling language to the cognitive conceptualizations of individuals involved in model creation and use [3, p. 130]. Nielsen [34, p. 153] refers to it as "speaking the user's language" and emphasizes the benefits of using domain-specific modeling concepts. Secondly, the formal cause can be related to the used concept specification mechanism (e.g., meta models, domain ontologies, or grammars). The type of concept specification mechanism is hypothesized to influence the understandability of the approach, e.g., depending on the number of concepts that are defined, the representation of specified concepts, and the offered descriptions [3,28]. Thirdly, the concrete syntax can be considered. To facilitate the understandability of model representations, it has been suggested that the concrete syntax should be tailored to the needs of users and their cognition [4,34,35]. An example catalog of principles that are assumed to promote the interpretation of graphical notational elements has been suggested by [4].

The *efficient cause* focuses on the creative act of constructing a model. Model creation is said to be influenced by the characteristics of involved actors (e.g., their abilities), the skills required to create a model, and the overall comprehensibility of the used modeling approach [6,7,36]. The 'comprehensibility' of a modeling approach, in turn, is in itself an intricate construct that has warranted specific investigations. For example, it has been found to affect the necessary learning efforts (i.e., the steepness of the learning curve) for both modelers and model users [6,34,36]. Lastly, the efficient cause can also be considered to encompass the 'productivity' of creating, analyzing, and modifying models. This again might depend on the modelers' skills and experiences, and may be further affected by the modeling language and available modeling tools (cf. [7, p. 10]).

The final cause can be related to the goals underlying the creation of a model. Every enterprise model is created under the assumption that it will be useful for particular purposes. These may include, e.g., creating a high-level common understanding, supporting analysis, or developing software [1,2]. In order to

Table 3. The 'self-explanatory capabilities'

Cause	Aspects	Comment
Material	The domain of discourse	How is the domain of discourse accessed?
Formal	Modeling language	Abstract and concrete syntax, semantics
Efficient	The act of model creation and required knowledge	Involved actors, required knowledge, learning efforts, procedure and productivity of modeling
Final	The end of model creation	Purposes of model use (e.g., analysis, implementation)

meet their assumed purposes, models might need to possess specific properties such as featuring a specific level of abstraction or a specific level of detail.

4   Comparative Analysis

In this section, we demonstrate the applicability of the framework by using it to characterize three selected EM approaches: ArchiMate [37], Design and Engineering Methodology for Organizations (DEMO) [38], and Multi-Perspective Enterprise Modeling (MEMO) [1]. These approaches were selected as they have been developed in interrelated, but not identical contexts, and thus, are particularly distinctive (cf. [2]). On a high level, DEMO is linked to the intention to capture only the 'essence' of organizational business processes, MEMO aims to provide comprehensive reconstructions of technical languages, while ArchiMate is tailored towards use with common EA frameworks.

4.1   'Causes' of an Enterprise

To identify the causes of an enterprise addressed by the approaches, we analyze their modeling concepts, and assess to what extent they directly or indirectly allow to express aspects related to a given cause. To this aim, we analyze Archi-Mate's specification defined in [37], DEMO's concepts, attributes and relationships derived from [39], and finally, available domain-specific modeling languages (DSMLs) being part of MEMO (e.g., OrgML [40,41], GoalML [42], MetricML [43], ResML [44], and ITML [45]). As the number of concepts offered by the approaches is altogether quite high, a detailed enumeration is not in the scope of this paper. Instead, we discuss the most important findings (see also Table 4).

Material cause. A clear distinction between the approaches can be detected: DEMO aims at separating "the essential issues from their realization" [46, p. 323], and therefore favors concepts describing speech acts instead of concepts of material and immaterial resources [38, pp. 81–86]. It is suggested that a set of basic

Table 4. Explanatory capabilities of selected approaches

Cause	Aspects	ArchiMate	DEMO	MEMO
Material	Elementary constituents	○	◑	○
	Resources	◑	○	●
	Personnel, skills and competencies	○	○	◑
	IT Infrastructure	●	○	●
Formal	Organizational structure, hierarchies	◑	◑	●
	Dynamic abstractions	◑	●	●
Efficient	Actual work being done	○	○	◑
	Metrics and indicators	○	○	●
Final	Goals, values and mission	◑	○	●

Legend: ○ = not covered; ◑ = partly covered; ● = largely covered types of speech acts (including, e.g., 'request', 'promise', and 'accept') can be used to describe generic 'transaction' patterns, based on which, in turn, business processes can be described from an elementary, 'essential' point of view [46, pp. 311–315]. It is a distinct property of DEMO that it conceives of the 'matter' of organizational processes as made up solely of communication. Consequently, DEMO exclusively suggests a basic set of concepts for describing communicational acts that can be found in (routine) business processes in an organization. In contrast, ArchiMate and MEMO offer concepts for describing material and immaterial resources. MEMO contains dedicated DSMLs accounting for resources in general [44] as well as IT resources in particular [45]. Archi-Mate mostly concentrates on IT infrastructures, which are construed as one 'layer' of an enterprise [37, p. 63]. In terms of the number of modeling concepts, attributes, and syntactical constraints, the MEMO specifications are semantically richer compared to those offered by ArchiMate or DEMO (cf. [2]).

The *formal cause* involves different aspects. Regarding the (formal) organizational structure, MEMO offers a dedicated DSML [41]. It includes a range of concepts adhering to the traditional organization theory (e.g., *OrganizationalUnit*, *Position, Board, Committee, Role*) and a number of specific relationships and constraints aiding the modeling of semantically correct organizational structures. Similarly, but less comprehensively, ArchiMate offers a selection of rather coarse concepts (e.g., *Business Actor, Business Role*) and generic associations (e.g., *'aggregation'*, *'assignment'*). No specific constraints are defined [37, pp. 18–47]. In contrast, DEMO focuses on the concept of an actor role (*Elementary Actor Role, Composite Actor Role*), which is an abstraction of individuals who have the authority to perform certain speech act patterns (i.e., transactions) on behalf of an organization. It is possible to define the scope of transactions, for which an actor is responsible (responsibility areas), but organizational structures in a conventional sense are not covered.

Further, when considering dynamic abstractions, each approach offers concepts for describing business processes, albeit in different ways. MEMO offers a dedicated DSML for describing business processes [40], which provides conventional concepts used in this context (e.g., *ControlFlowSubProcess, Event*) as well as more elaborate concepts (e.g., *Exception*). ArchiMate provides a limited number of concepts (*Business Process, Business Event*) for which no syntactical constraints are specified. DEMO's sole purpose is describing business processes, though it does so through the lens of speech acts (see above). It follows that the concepts employed by DEMO (e.g., *Coordination Act, Production Act, Coordination Fact*) are largely different from those commonly used in business process modeling. The approaches also offer concepts for describing broader activity or responsibility areas, which are neither purely structural, nor purely sequentially defined. ArchiMate, to this end, provides the concept *'Business Function'* [37, pp. 30–31]. MEMO OrgML offers a concept *'task'* [41, p. 50] to model work packages that are possibly part of a process. Finally, with respect to the informal organization, there are no explicit concepts provided by the different languages.

Efficient Cause. To our best knowledge, neither DEMO nor ArchiMate provides direct means to capture the actual work done. In contrast, MEMO offers conceptual means to track concrete values for activity-related concepts such as *Process* at the instance level (e.g., '*startTime*' and '*stopTime*'; specified as 'intrinsic' attributes). Based on that, attributes are offered to aggregate the data at type level (e.g., average process duration). When it comes to metrics and indicators, again, MEMO is the only approach providing concepts to explicitly describe performance measures (e.g., *Indicator*, being part of MetricML [43]).

The *final cause* is explicitly addressed only by ArchiMate and MEMO. DEMO does not offer specific concepts for describing aspirational or value aspects. ArchiMate offers concepts such as *Value* and *Goal* [37, pp. 18–47, 141–168], but the specification remains at a rather basic level of semantics (there are no attributes, syntactical constraints; and only the above-indicated generic relationships are available). MEMO offers a dedicated DSML, GoalML, which defines a more comprehensive set of concepts allowing to express enterprise goals (e.g., *EngagementGoal, SymbolicGoal, GoalConfiguration* [42]).

4.2   Self-explanatory Capabilities

The material cause. In order to create an enterprise model, all considered approaches suggest to collaborate with enterprise key members, and to analyze existing documents. Thus, to some extent, each approach builds on the account provided by organization members to construct models. But there are different points of emphasis. The basis for specifying DEMO models is a rather strict (textual) analysis of "all available documentation about the enterprise" [38, pp. 142–143]. This analysis is intended to yield candidates for 'actors' as well as various kinds of their 'acts' and 'results', from which more comprehensive transaction patterns can be constructed [38, pp. 142–154]. Although it is noted that such an analysis typically involves interpretations [38, p. 143], DEMO claims that modeling results in "a correct and complete set of [...] models of an enterprise ontology" [38, p. 142]. Hence, existing (textual) accounts serve to rather rigorously identify a supposed factual 'essence' of the status quo. The application of all DSMLs being part of the MEMO method is guided by their own process models [1, p. 951]. Despite domain-specific differences it is usually emphasized that there are different ways to construct a model, and that the scope of a resulting model may depend on the given use scenario (e.g., [45, pp. 384–422]). In contrast to DEMO, there is no assumption that modeling needs to result in one "correct and complete" model. MEMO is linked to the intention to 'empower' domain stakeholders by offering a modeling language, which suits their world of expertise [1, p. 946]. Hence, the 'material' from which a MEMO model is built stems to a larger extent from deliberate reflection on the status quo, future aims, and the case-specific purpose of a model. ArchiMate itself does not specify how the information on the domain required to create a model is to be gathered, but it can be used in conjunction with EA framworks such as TOGAF [37, pp. 14–16]. TOGAF includes a comprehensive procedural description drawing on existing documents and stakeholder communication [47, pp. 43–194].

The Formal Cause. On a general level, the analyzed approaches vary significantly with respect to their *conceptual scope* and the *closeness* of the modeling concepts to the perception of prospective users. ArchiMate and MEMO offer concepts that are intendedly resembling (but not necessarily identical to) professional language concepts of prospective users. DEMO builds on a markedly different view of organizations, resulting in an idiosyncratic set of concepts. MEMO is grounded on the explicit assumption that offering modeling concepts, which reconstruct the technical languages of certain actor groups (i.e., of certain professional domains), is useful because "it will reflect characteristic goals, common practices and preferred levels of abstraction" [1, p. 945]. It thus offers semantically more elaborate DSMLs for various domains that include a large number of concepts and syntactical constraints (cf. [2]). At first sight, ArchiMate follows a similar intention: It also aims to offer domain-specific concepts for covering enterprise domains ranging from a 'business' layer to an 'IT' layer [37, pp. 3–10]. But the central assumption of ArchiMate is that a limited set of modelling concepts is both sufficient and helpful for the modelling of enterprise architectures [37, p. 2]. In consequence, the ArchiMate modeling concepts are more coarse, and the overall language design is rather flexible in the sense that few syntactical constraints are available, and most relationships remain generic (cf. [2]). In contrast, DEMO does not orient itself towards existing language structures of the (professional) application domains but builds on an individual view of operational processes in organizations (self-labeled "PSI theory", and defined in terms of a set of "axioms"; see [38, pp. 81–114]). This view is inspired by selected insights from communication research and the philosophy of language. Accordingly, it is assumed that elementary, communicational constituents of operational processes can be identified in actual enterprises, that these elements can be used to capture the 'essence' of these organizational processes, and, finally, that this way of viewing business processes is more "effective" (in some sense) than using conventional concepts (cf. [38, pp. 7–13]). In other words, DEMO does not reconstruct the user's language, but offers its own, arguing that it is better suited to achieve the stated goals. With regard to the *concept specification* (i.e., the means of specifying the abstract syntax), only a few comments are relevant here (for a more detailed discussion, see [2]). The abstract syntax of ArchiMate is defined by meta models and tables but there is no explicit meta modeling language, and there are almost no further syntactical constraints. MEMO offers comprehensive DSMLs specified by meta models defined in a common meta modeling language, such as to form an integrated language architecture [1, p. 946]. DEMO provides a formal description of modeling concepts, however, to our best knowledge, no meta model is available yet. With regard to the *model representation*, MEMO can be found to be most closely oriented towards the human cognition as it emphasizes several notational guidelines as suggested by, e.g., [4]. Only some ArchiMate symbols are associatively linked to the concepts they represent. DEMO symbols do not directly convey indications about their semantics, i.e., "their meaning is purely conventional and must be learnt" [4, p. 764]. Finally, ArchiMate and MEMO offer mechanisms to deal with model complexity, using 'viewpoints' [37, pp. 97–135] or perspective-specific diagram types [1].

The Efficient Cause. DEMO presupposes that users are familiar with the particular theoretical view underlying the offered modeling concepts. In consequence, learning efforts will usually be required in order to get to know and master the DEMO approach [48]. DEMO users need to be familiar with a particular way of seeing their area of concern. On the contrary, MEMO assumes that the developed enterprise models are meaningful on their own and that only domain-knowledge is required in order to use the given approach. Thus, MEMO aims at empowering users, i.e., enabling them to be modelers. Notwithstanding this intention, it can be assumed that many conceptual definitions of MEMO, and especially sophisticated specifications regarding different levels of abstractions, go well beyond everyday conceptualizations of targeted users (see, e.g., the language design arguments raised by [42, pp. 116, 196–198]). The act of modeling in the case of ArchiMate is performed by enterprise architects, who are expected to be familiar with the concepts from the EA domain. As the concrete syntax is less tailored towards the human cognition (in terms of the guidelines by [4]), ArchiMate models may be visually less intuitive than MEMO models.

The final cause. Each approach defines activities that should be enabled or supported by the created models. ArchiMate models are intended to meet generic conceptual descriptive needs of EA frameworks. DEMO models are supposed to enable the analysis and design of elementary process-oriented aspects of enterprises and thus to 'engineer' an enterprise [38, p. 74]. MEMO models rather aim to support discursive analyses and redesigns related to various organizational domains, while also being transformable into implementation level artifacts [1].

4.3 Discussion

Models, following a pragmatic view, can be seen as 'means-to-an-end', i.e., "particular instruments that should provide some value when used for the intended goals by the intended users" [19, p. 433]. Thus, the scope of the provided modeling concepts (the first explanatory capability) as well as the understandability of a modeling approach and the resulting models (self-explanatory capability) should correspond to the stated goals of the given approach.

Not surprisingly, the *explanatory capabilities* of the approaches vary significantly. DEMO focuses on the material and formal causes as, according to its main assumptions, parts of these reflect the essence of an organization. ArchiMate covers selected aspects of the material, formal and final cause, in line with the considered general application scenarios in the EA domain. MEMO, by intending to comprehensively reconstruct domain languages, tends to cover more aspects (and hence, comprises more modeling concepts) compared to the other approaches. The explanatory capabilities of all analyzed approaches—however different they may be—are in line with their stated assumptions and can indeed be assumed to support the realization of the stated goals. The *self-explanatory capabilities* of the analyzed enterprise modeling approaches vary as well. A clear advantage of DEMO is the fact that it permits to abstract from a significant share of the operational details of activities involved in an organizational business process. However, due to the narrow focus, the high level of abstraction, and the unconventionality of the approach, it can be presumed that DEMO is not immediately understandable to most domain experts and users [48]. It may be for this reason that training and certification possibilities are offered by its creators. In addition, the graphical representation of DEMO models does not seem to be particularly suited to the human cognition. Apparently, the concrete syntax has not been at the center of attention yet. In contrast, ArchiMate provides more domain-specific concepts. Although these concepts are rather coarse and generic, they might suit the needs of common EA practices. Finally, the modeling concepts provided in the MEMO DSMLs aim to feature a high degree of correspondence with domain-specific terminologies. For those concepts, which meet this aim, it can be assumed that there is a high degree of understandability by domain experts. This promises to offer an instrument that is directly oriented to the 'worldview' of the actors whom it is supposed to support. However, at the same time, several MEMO concepts are specified at a level of elaborateness which will certainly require non-negligible learning efforts. The graphical representation of MEMO modeling concepts is mostly oriented towards visualizations originating from the domain. In sum, MEMO might be seen as being closest to the human conceptualization of the domain of discourse, as well as to the human cognition with respect to the model representation. However, when compared to ArchiMate, the learning effort and the modeling skills demanded by MEMO are presumably higher.

5 Conclusions

In this paper, drawing on the four causes of Aristotle, we designed a framework that can be used to investigate the explanatory capabilities of EM approaches. The performed analysis of three concrete approaches indicated both strengths and shortcomings of the framework. An important benefit of the framework is that it offers a set of categories which seems to be abstract enough to allow for a characterization of very diverse modeling approaches. This may serve as a basis for their further comparison. However, a shortcoming of the framework is that it does not yet include strictly clear-cut metrics that would allow for a more in-depth and 'objective' comparison of the specific features, especially when it comes to the self-explanatory dimension. Finally, the ambiguity of the Aristotelian causes in the light of modern science has to be acknowledged. Nevertheless, we argue that the application of the framework, with the new interpretations proposed, can contribute to the further evolution and integration of modeling approaches. To this end, various paths for future research can be considered. On the one hand, we aim to investigate ways to operationalize the introduced categories of the framework. On the other hand, we seek to further validate the framework by analyzing a wider set of modeling approaches.

References

1. Frank, U.: Multi-perspective enterprise modeling: foundational concepts, prospects and future research challenges. SoSym **13**(3), 941–962 (2014)
2. Bock, A., Kaczmarek, M., Overbeek, S., Heß, M.: A comparative analysis of selected enterprise modeling approaches. In: Frank, U., Loucopoulos, P., Pastor, Ó., Petrounias, I. (eds.) PoEM 2014. LNBIP, vol. 197, pp. 148–163. Springer, Heidelberg (2014)
3. Kirikova, M.: Explanatory capability of enterprise models. Data Knowl. Eng. **33**(2), 119–136 (2000)
4. Moody, D.L.: The physics of notations: toward a scientific basis for constructing visual notations in software engineering. IEEE TSE **35**(6), 756–779 (2009)
5. Krogstie, J.: Modelling of the people, by the people, for the people. In: Krogstie, J., Opdahl, A., Brinkkemper, S. (eds.) Conceptual Modelling in Information Systems Engineering, pp. 305–318. Springer, Berlin (2007)
6. Lankhorst, M.: Enterprise Architecture at Work: Modelling, Communication and Analysis. The Enterprise Engineering Series, 3rd edn. Springer, Heidelberg (2013)
7. Frank, U.: Power-modelling: toward a more versatile approach to creating and using conceptual models. In: Proceedings of 4th International Symposium on BMSD, pp. 9–19 (2014)
8. Siau, K., Rossi, M.: Evaluation techniques for systems analysis and design modelling methods - a review and comparative analysis. ISJ **21**(3), 249–268 (2011)
9. Strahringer, S.: Metamodellierung als Instrument des Methodenergleichs. Evaluierung am Beispiel objektorientierter Analysemethoden. Shaker, Aachen (1996)
10. Stachowiak, H.: Allgemeine Modelltheorie. Springer, Wien (1973)
11. Johnson, P., Ekstedt, M.: Enterprise Architecture: Models and Analyses for Information Systems Decision Making. Lightning Source Incorporated (2007)
12. Cilliers, P.: Making sense of a complex world. In: Aaltonen, M. (ed.) The Third Lens: Multi-Ontology Sense-Making & Strategic Decision-Making, pp. 99–110. Ashgate, Hampshire (2007)
13. McIntyre, L.: Complexity: a philosopher's reflections. Complex. **3**(6), 26–32 (1998)
14. Frank, U.: Multi-perspective enterprise modelling: background and terminological foundation. ICB-Research Report 46, Universität Duisburg-Essen (2011)
15. Rumbaugh, J.: Objects in the constitution - enterprise modeling. J. Object-Oriented Program. **5**(8), 18–24 (1993)
16. Leist-Galanos, S.: Methoden zur Unternehmensmodellierung. Vergleich, Anwendungen und Integrationspotentiale. Logos, Berlin (2006)
17. Buckl, S., Schweda, C.M.: On the state-of-the-art in enterprise architecture management literature. Technical report, sebis, Technical University Munich (2011)
18. Bork, D., Fill, H.G.: Formal aspects of enterprise modeling methods: a comparison framework. In: Proceedings of HICSS-47, pp. 3400–3409 (2014)
19. Bjeković, M., Proper, H.A., Sottet, J.-S.: Embracing pragmatics. In: Yu, E., Dobbie, G., Jarke, M., Purao, S. (eds.) ER 2014. LNCS, vol. 8824, pp. 431–444. Springer, Heidelberg (2014)
20. Juarrero, A.: Causality and explanation. In: Allen, P., Maguire, S., McKelvey, B. (eds.) The Sage Handbook of Complexity and Management, pp. 155–164. SAGE Publications Ltd, Thousand Oaks (2011)
21. McKelvey, B.: Toward a complexity science of entrepreneurship. J. Bus. Ventur. **19**(3), 313–341 (2004)

22. Kaminska-Labbe, R., McKelvey, B., Thomas, C.: On the coevolution of causality: a study of aristotelian causes & other entangled influences. Working paper, CERAM, Sophia Antipolis, France (2008)
23. Kosiol, E.: Organisation der Unternehmung. Gabler, Wiesbaden (1962)
24. March, J.G., Simon, H.A.: Organizations. Wiley, New York (1958)
25. Cyert, R.M., March, J.G.: A Behavioral Theory of the Firm. Prentice-Hall, Englewood Cliffs (1963)
26. Taylor, J.R.: Rethinking the Theory of Organizational Communication: How to Read an Organization. Ablex, Norwood (1993)
27. Figl, K., Laue, R.: Cognitive complexity in business process modeling. In: Mouratidis, H., Rolland, C. (eds.) CAiSE 2011. LNCS, vol. 6741, pp. 452–466. Springer, Heidelberg (2011)
28. Sarshar, K., Loos, P.: Comparing the control-flow of EPC and Petri Net from the end-user perspective. In: van der Aalst, W.M.P., Benatallah, B., Casati, F., Curbera, F. (eds.) BPM 2005. LNCS, vol. 3649, pp. 434–439. Springer, Heidelberg (2005)
29. Dahlbom, B., Mandahl, M.: A theory of information technology use. In Kerola, I.P. (ed.) Precedings of the 17th Information Systems Research Seminar in Scandinavia, pp. 66–67. University of Oulu (1994)
30. Barnes, J.: The Complete Works of Aristotle. Princeton University Press, Princeton (1993)
31. Gutenberg, E.: Die Unternehmung als Gegenstand betriebswirtschaftlicher Theorie: Unveränderter Nachdruck der Auflage Berlin 1929. Gabler, Wiesbaden (1998)
32. Grochla, E.: Einführung in die Organisationstheorie. Poeschel, Stuttgart (1978)
33. Cooren, F., Kuhn, T., Cornelissen, J.P., Clark, T.: Communication, organizing and organization. Organ. Stud. **32**(9), 1149–1170 (2011)
34. Nielsen, J.: Enhancing the explanatory power of usability heuristics. In: Proceedings of Conference on Human Factors in Computing Systems, pp. 152–158. ACM, New York (1994)
35. Mendling, J., Reijers, H.A., Cardoso, J.: What makes process models understandable? In: Alonso, G., Dadam, P., Rosemann, M. (eds.) BPM 2007. LNCS, vol. 4714, pp. 48–63. Springer, Heidelberg (2007)
36. Aranda, J., Ernst, N., Horkoff, J., Easterbrook, S.: A framework for empirical evaluation of model comprehensibility. In: Proceedings of the International Workshop on Modeling in Software Engineering, p. 7. IEEE Computer Society (2007)
37. The Open Group: ArchiMate 2.0 specification: Open Group Standard. The Open Group Series. Van Haren, Zaltbommel (2012)
38. Dietz, J.L.G.: Enterprise Ontology: Theory & Methodology. Springer, Berlin (2006)
39. Dietz, J.L.G.: Demo-3: Models and representations (version 3.7) (2014)
40. Frank, U.: MEMO Organisation Modelling Language (2): Focus on Business Processes. ICB Research Report 49, University of Duisburg-Essen (2011)
41. Frank, U.: MEMO Organisation Modelling Language (1): Focus on Organisational Structure. ICB-Research Report 48, University of Duisburg-Essen (2011)
42. Köhling, C.: Entwurf einer konzeptuellen Modellierungsmethode zur Unterstützung rationaler Zielplanungsprozesse in Unternehmen. Ph.D. thesis, University of Duisburg-Essen (2012)
43. Strecker, S., Frank, U., Heise, D., Kattenstroth, H.: MetricM: a modeling method in support of the reflective design and use of performance measurement systems. ISeB **10**(2), 241–276 (2012)
44. Jung, J.: Entwurf einer Sprache für die Modellierung von Ressourcen im Kontext der Geschäftsprozessmodellierung. Logos, Berlin (2007)

45. Heise, D.: Unternehmensmodell-basiertes IT-Kostenmanagement als Bestandteil eines integrativen IT-Controllings. Logos, Berlin (2013)
46. Dietz, J.L.G.: Demo: towards a discipline of organisation engineering. Eur. J. Oper. Res. **128**(2), 351–363 (2001)
47. The Open Group: TOGAF Version 9.1. Van Haren, Zaltbommel (2011)
48. Décosse, C., Molnar, W.A., Proper, H.A.: What does DEMO Do? A qualitative analysis about DEMO in practice: founders, modellers and beneficiaries. In: Aveiro, D., Tribolet, J., Gouveia, D. (eds.) EEWC 2014. LNBIP, vol. 174, pp. 16–30. Springer, Heidelberg (2014)

A Non-arbitrary Method for Estimating IT Business Function Recovery Complexity via Software Complexity

Athanasios Podaras[(⊠)]

Faculty of Economics, Department of Informatics,
Technical University of Liberec,
Studentská 1402/2, 461 17 Liberec 1, Czech Republic
athanasios.podaras@tul.cz

Abstract. The goal of the present paper is to introduce a new model for estimating business function recovery complexity in order to predict reasonable recovery timeframes in case of an unexpected information system failure. The method has its roots in the Use Case Points approach, which is a broadly tested tool for software complexity estimation. The current paper illustrates the pure theoretical form of the new model as well as the mapping between software complexity and business function recovery complexity. The method includes 3 categories of factors which affect the recovery procedure and are weighted according to the Rank Order Centroid (ROC) approach of assigning weights. The method is entitled Business Continuity Points. The idea behind the development of the new method is the establishment of a standard approach for implementing efficient time management regarding business function recovery. The estimated recovery time depends on the impact of technical, environmental and unexpected factors. Each function's Recovery Time should be compared with the Recovery Time Objective (RTO) and Maximum Accepted Outage (MAO) values as they are proposed by business continuity and IT experts.

Keywords: Use case points · Business continuity points · Business function recovery · Complexity estimation

1 Introduction

In modern information age, organizations are required to use IT in order to maintain the business operations and keep their competitive advantage in the market [1]. As a consequence, speedy restoration of services for critical organizational processes in the event that there are operational failures due to natural or man-made disasters [2] is imperatively demanded. Business continuity focuses on ensuring an organization can continue to provide services when faced with various crisis events [3]. The current work delineates a theoretical model which aims to assist Business Continuity Managers in managing business function recovery time, via a business process complexity estimation method. The method is entitled *Business Continuity Points*. It is aimed to recover a singular business process or an entire business function which is divided into several processes. The method finds its roots in the Use Case Points [4] approach to the

© Springer International Publishing Switzerland 2015
D. Aveiro et al. (Eds.): EEWC 2015, LNBIP 211, pp. 144–159, 2015.
DOI: 10.1007/978-3-319-19297-0_10

estimation of Software Complexity. The idea behind the construction of the new method, is that by following the rules of software development complexity, we can define the recovery complexity of a specific business function or process, and consequently of its IT infrastructure in case of an unexpected interruption. A mapping between the two approaches is presented for the better understanding of the presented approach.

Through the estimation of recovery complexity, the recovery time of a business function may also be predicted. The recovery time calculation can assist IT managers and business continuity experts in defining more precise Recovery Time Objective (RTO) and the Maximum Accepted Outage (MAO) timeframes [5], which are an indispensable part of the Business Impact Analysis. Business Impact Analysis (BIA) [6, 7] helps develop business recovery objectives by determining how disruptions affect various organizational activities. BIA seeks to quantify the impact of possible events and provides the foundation for developing continuity and recovery strategies [3]. The prediction of the approximate recovery time should be based on simple, average and complex recovery scenarios, considering the severity of the factors which influence the recovery procedure.

2 Problem Definition

In the occasion of an unexpected failover of an information system, the business functions, the involved processes which are parts of the function and the technical infrastructure, should be recovered within a reasonable and acceptable timeframe, so that the enterprise will not suffer an irrevocable financial loss. This timeframe is indicated by the Rational Time Objective (RTO) and the Maximum Accepted Outage (MAO) values. The former value refers to a reasonable recovery time required to recover a business function or process, while the latter expresses the maximum tolerable time of a system's interruption, the surpassing of which will result to a significant financial loss. The accepted downtime period is determined in terms of a cautiously formulated business continuity policy. Darril Gibson [8] indicates the impact value level of each business function according to its accepted downtime period. The four levels of impact value are:

Level 1: The business function should operate without any interruption. Online systems must be available 24 h per day and 7 days per week.

Maximum Acceptable Outage (MAO) = 2 h
Recovery Time Objective (RTO) < 2 h

Level 2: The business processes can survive without the business function for a short amount of time.

Maximum Acceptable Outage (MAO) = 24 h (1 day)
Recovery Time Objective (RTO) < 24 h

Level 3: The business processes can survive without the business function for one or more days.

Maximum Acceptable Outage (MAO) = 72 h (3 days)
Recovery Time Objective (RTO) < 72 h

Level 4: The business processes can survive without the business function for extended periods.

Maximum Acceptable Outage (MAO) = 168 h (1 week)
Recovery Time Objective (RTO) < 168 h

According to Gibson, who is an expert in Business Continuity Management, the above estimated values are internal, which means that recovery objectives used by one organization can be completely different from those used by another organization [8]. Nevertheless, though flexible, the above values are considered as reliable, due to the fact that both direct and indirect costs have been considered for their calculation. Direct costs include, i.e. loss of immediate sales and cash flow or equipment/building replacement costs, while indirect costs include i.e. lost opportunities during recovery. In addition, the validity of the above stated values can be controlled in comparison to similar estimated values by the National Institute of Standards and Technology – U.S. Department of Commerce [9], where specific business tasks and the involved systems have been assigned the corresponding RTO and MAO values according to the experts of the Institute.

However, the efficient estimation of RTO and MAO values is still a very big issue for enterprises nowadays. The author of the current work, having years of practical experience in the IT department of a bank, has concluded that the determination of RTO and MAO is done in a non-objective manner, without using a standard method, and it is based only on the everyday experience of the managers, which almost always leads to a rejection by the business continuity team. Consequently, valuable time is lost due to multiple repetitions of the business continuity policy formulation. Moreover, multiple experts underline the arbitrary assignment of the recovery time estimation. Snedaker [10] indicates that HR might say "we have to have our payroll application"; marketing might say "without our CRM system, we can't sell any products"; manufacturing might say "without our automated inventory management system, we can't even begin to make anything." Therefore, the IT department's critical business functions are driven externally, to a large degree.

Considering the aforementioned enterprise reality, the author was inspired to develop and propose a standard tool for defining more precise business continuity timeframes. The timeframes (RTO and MAO) are recorded in the Business Impact Analysis document, which is a Requirement Analysis document. Consequently, the new method needs to include a practically tested requirement analysis tool, which estimates both system complexity and system development time. The selected tool is the Use Case Points approach, which deals with all the aforementioned issues. Furthermore, flexibility issues forced the author of the current work to design a general model for business process recovery complexity, and avoid listing specific factors, as in the Use Case Points approach, which may limit its practical value. For the same reason the weights of the factors should not be arbitrarily assigned, but be calculated by a standard mathematical model. The model applied for this purpose is the Rank Order Centroid (ROC) method for assigning weight values.

3 The Business Continuity Points Method

The currently proposed method is based on the Use Case Points approach to the estimation of software complexity [11–13]. It is derived from the UML Use Case model which aims to thorough requirement analysis. In order to avoid repetition and due to the rich available literature around Use Case Points, the analysis of the method is not included in the objectives of the current paper. However, critical points of the approach are depicted in the mapping of the Use Case Points and the Business Continuity Points method (Table 2). The present section includes a delineation of the Business Continuity Points method. The overall business function recovery complexity estimation model is depicted in Fig. 1.

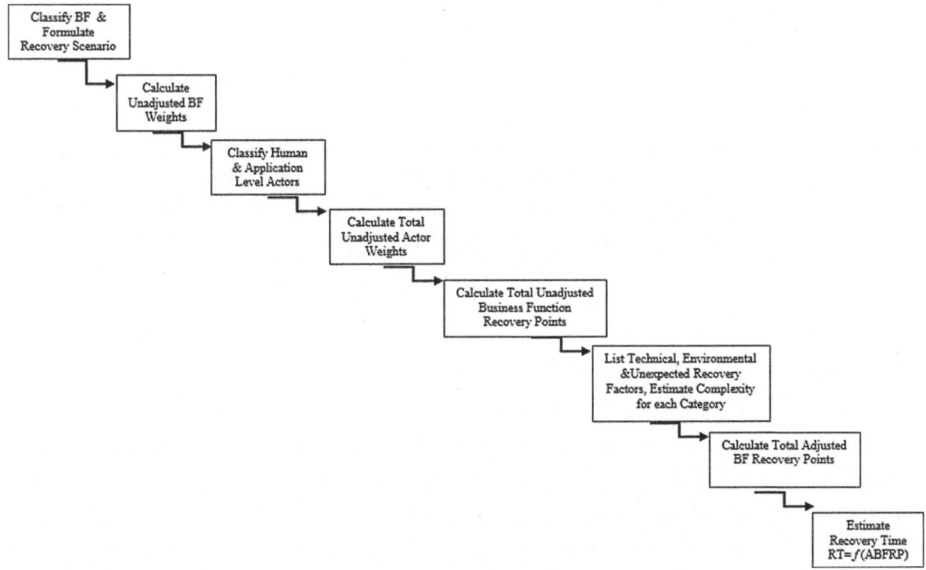

Fig. 1. Business function recovery complexity for the estimation of approximate recovery time

3.1 Classification of Business Functions

The first part of the method refers to the classification of business function types. The differentiation from the Use Case Points method is that it classifies Use Cases according to the number of transactions involved in each Use Case, whereas the new method, since it focuses on business function recovery complexity, implements Business Function Classification according to the number of *Processes* included in the function.[1] More precisely, business functions are classified as:

---

[1] Classification can also be performed for a single business process according to the number of *Activities* inside the process.

Simple Business Function in which Number of Business Processes is <=3,
Average Business Function in which Number of Business Processes is >= 4
    and <=7,
Complex Business Function in which Number of Business Processes is >7.

The corresponding Weighting Values of the Business Function Types are 0.5, 1 and 1.5. The specific values were selected in order to derive reasonable estimations of Recovery Time, which should be in accordance with the Rational Time Objective (RTO) and Maximum Accepted Outage (MAO) values that are proposed by Gibson [8]. Yet, these values can be further modified since RTO and MAO can be also altered to meet the needs of a specific enterprise and its particular classification of its business functions. The calculation of Unadjusted Business Process Weights (UBFW), is performed via the following equation:

$$\text{UBFW} = \sum_{i=1}^{n} (\text{BF}_i \times \text{W}_i) \tag{1}$$

where, n is the Number of Business Processes, BF_i is the Type of the given Business Function i and W_i is the Weight of the corresponding Business Function.

The score is obtained by counting the number of Business Functions of each type (complex, average or simple), multiplying each total by its weighting factor and adding up the products.

3.2   Classification of Actors and Calculation of Unadjusted Business Function Recovery Points

Actors, in the Use Case Points method, are distinguished, in Simple, Average and Complex [11]. A Simple Actor represents another (or external) system with a defined Application Programming Interface, API, an Average Actor is another system interacting through a protocol such as TCP/IP, and a Complex Actor may be a person interacting through a GUI or a Web page. The corresponding Weighting Values of the Actors are 1, 2 and 3.

However, a more detailed Actor classification is implemented in the Business Continuity Points. Due to the fact that business function recovery is under human as well as software application influence, the Actors are classified in Actor Type 1 – A1 (Human Level Actors) and Actor Type 2 – A2 (Application Level Actors). Application Level Actors are classified in the same way as in Use Case Points with a difference in weight values. That is, 1.5 (Complex), 1 (Average) and 0.5 (Simple). Human Level Actors are classified in the following way:

Level 1 (Complex): IT Managers on top of the IT Department or a corresponding division in a company who are leaders of the Business Continuity team. The weight value assigned to the personnel of this level is 1.5.
Level 2 (Average): IT subdirectors or supervisors of an IT Section who are members of the recovery team. The weight value assigned to the personnel of this level is 1.

Level 3 (Simple): Employees of an IT Department who are members of the recovery team. The weight value assigned to the personnel of this level is 0.5.

The difference in weight values is attributed to the achievement of reasonable recovery time effort results, according to the Business Standard Institute. The values were determined after the appropriate calculations were made by the author.

The Equations utilized for Unadjusted Weights are the following:

$$UHW = \sum_{i=1}^{n} (A1_i \times W_i) \tag{2}$$

where UHW is the Unadjusted Human Weight value, $A1_i$ is Human Level Actor i, and W_i is the Actor's Weight, for Human Level Actors, and similarly,

$$UAPW = \sum_{i=1}^{n} (A2_i \times W_i) \tag{3}$$

where UAPW is the Unadjusted Application Weight value, $A2_i$ is Actor i, and W_i is the Actor's Weight, for Application Level Actors.

The total score of Unadjusted Actors' Weights TUAW is provided by the formula:

$$TUAW = UHW + UAPW \tag{4}$$

The score is obtained by counting the number of Actors of each type (complex, average or simple), multiplying each total by its weighting factor, adding up the products and adding the UHW and UAPW values. Finally, the Unadjusted Business Function Recovery Points (UBFRP) value is calculated according to the following formula:

$$UBFRP = TUAW + UBFW \tag{5}$$

3.3   Classification of Factors

The second part of the model includes the creation of 3 categories of factors which may have a strong influence on the business function recovery task. The specific factors are distinguished in *Technical Recovery Factors, Environmental Recovery Factors* and finally *Unexpected Recovery Factors*. The new element which reveals the method's differentiation from the Use Case Points model is the last category. It was inspired due to the fact that in the occurrence of a real information system failure, unexpected events may seriously delay the recovery procedure. Consequently, apart from technical and environmental factors, which are also included in the Use Case Points and can be foreseen by business continuity managers, unexpected factors should be also taken into account throughout the formulation of an efficient business continuity plan. However, in order to ensure the flexibility and expandability of the model, the creation of lists with specific factors is avoided. Instead, the paper focuses on the proposal of these 3 categories of factors and allows business continuity managers to create their own list of factors and rank them according to their personal experience. Furthermore, they can rank factors according to the importance and type of a given business function. In order to clarify the role of the factors, the IT business continuity managers should keep in mind a short delineation for each category.

Technical Recovery Factors (TRFs) mainly relate to the influence of the Technical Entities, which are involved in the business function, on the recovery process which should be recovered after outage. Technical factors refer to applications, platforms, interfaces, hardware and network components which are related to the business process or function. Examples of Technical Recovery Factors are Application's communication or dependency on other systems/applications, Business Function Type, Security Features and many more.

Environmental Recovery Factors (ERFs) mainly relate to the effect of Human Entities and their behavior on the recovery process. Human Entities can be users, business experts, a business recovery team, business owners, consultants and many other people who are responsible for the operation of the business function or process. Examples of Environmental Recovery Factors are the User's Experience, the Business Process Recovery Knowledge of the Team and other factors inspired by the Business Continuity Management.

Unexpected Recovery Factors (URFs). The existence of this category is considered to be indispensable, due to the business continuity and system recovery concept. URFs mainly relate to unplanned and unpredictable situations and scenarios that may emerge during the recovery process of a business function, and may significantly delay the process by exceeding the RTO and MAO values. Unexpected Factors can be Staff Availability, Network Availability, Disaster Type, Timely Information Distribution and many more. The term "unexpected" indicates that the emergence of these factors during the recovery process is unpredictable, and so is their precise impact. The flexibility of the current method allows for the use of an unlimited number of factors as well the simultaneous change of their corresponding weight values, ensuring that IT Business Continuity Managers can continually update the list of unexpected factors based on both events happening to the specific enterprise or others around the world.

Similarly to the Use Case Points theory, a calculation of the total value for each of the above factors should be implemented. The estimation of these values is dependent on the *Weight*, and *Assessment Value* of each factor.

3.4 Assigning Weights According to the Rank Order Centroid (ROC) Method

In order to avoid the method's limited functionality, an unlimited number of factors can be documented for each category, contrary to the Use Case Points Method which proposes only 13 Technical and 8 Environmental factors that influence the software development process. The unlimited number of factors in the presented model, triggers the necessity of assigning non-arbitrary weight values for each factor. The selected approach for assigning weights in the present method is the Rank Order Centroid (ROC) [14] approach.

The primary reason for selecting the ROC approach is that it is a simple way of giving weight to a number of items ranked according to their importance. The decision makers can usually rank items much more easily than give weight to them. This method takes those ranks as inputs and converts them to weights for each of the items, according to the following formula [15]:

$$W_i = \frac{1}{m} \times \sum_{n=1}^{m} \frac{1}{n} \tag{6}$$

where W_i is the Weight Value of the ith item, and m denotes the number of items (factors). Furthermore, the ROC approach enhances the flexibility of the entire Business Continuity Points method. Since the sum of the weight values must be always equal to 1, no matter what the number and the decided ranking order of the factors is, the estimated effort required to recover the business function is not affected. It can be thus stated that the model can be adjusted to the needs of every business continuity plan, and its limited functionality is obviously avoided. For deriving a reasonable recovery complexity estimation, the assigned weight values should be multiplied by 10. As a result the final weight values are normalized on a 0 to 10 interval scale.

3.5   Determination of Assessment Values for Each Factor

Another important element of each factor is its Assessment Value. Assessment Value indicates the severity of the factor and its impact on the recovery procedure. The determined Assessment Values of each factor are assigned according to a 4-level scale. The minimum value is 1 and the maximum value is 4. The scale can be either ascending or descending. Moreover, it can be Boolean or non- Boolean.

Type 1: *Factors with an Ascending Scale of Assessment Values:* In this category, according to the model, the higher the assessment value of the factor, the higher the degree of influence that the factor has on the recovery process. The factors with a low level of assessment value are marked with 1 and the factors with the highest influence on the recovery process are marked with 4. The factors are evaluated according to either a 4-level scale or a 2-level scale. The Scale is determined according to the type of the considered factor. The type of factors can be also Boolean (YES/NO, i.e. Exists Backup Site) or non-Boolean (i.e. Functional/Business Area (Importance/Criticality). The former require 2-level-scale assessment values while the latter require 4-level-scale. Boolean type factors indicate either positive or negative effects on the business function's recovery procedure. Thus, the existence of intermediate assessment values is avoided by the author.

Type 2: *Factors with a Descending 4-level Scale of Assessment Values:* In this category, according to the model, the higher the value of the factor, the lower the degree of influence that the factor has on the recovery process (i.e. Easy to Process Application/ System) (Table 1).

Table 1. Assessment values for technical, environmental and unexpected factors.

Assessment value	Boolean (YES/NO)	Non-boolean
Ascending scale	1, 4	1, 2, 3, 4
Descending scale	4, 1	4, 3, 2, 1

3.6 Derivation of TRF, ERF and URF Formulas and Adjusted Business Function Recovery Points

The derivation of formulas which should calculate the Technical, Environmental and also Unexpected Recovery Factors, is a crucial part of the current work. The formula, which provides the average TRF value, which is similar to the equation which provides the value of the Technical Complexity Factor (TCF) in the Use Case Points model, is the following:

$$TRF = c_1 + \frac{1}{c_2} \times \sum_{i=1}^{n} \left(\frac{Fi_{max} + Fi_{min}}{2} \times W_i \right) \tag{7}$$

where, $Fi_{max} = 4$ and $Fi_{min} = 1$ are the maximum and minimum assessment values of a recovery factor i, W_i is the Weight Value of the specific factor, n is the number of the determined factors, c2 is the speed of increasing recovery complexity and c1 is a correcting constant.

According to the Use Case Points model, average recovery complexity should be equal to 1. Therefore, the above equation can be written:

$$1 = c_1 + \frac{1}{c_2} \times \sum_{i=1}^{n} \left(\frac{Fi_{max} + Fi_{min}}{2} \times W_i \right) \tag{8}$$

The above equation can also be written in the following way:

$$c_1 = 1 - \frac{1}{c_2} \times \sum_{i=1}^{n} \left(\frac{Fi_{max} + Fi_{min}}{2} \times W_i \right) \tag{9}$$

If the speed of increasing recovery complexity is the same as the technical complexity in the Use Case Points method, we can compute c2 value (c2 = 100) and get the following formula:

$$TRF = 0.75 + \frac{1}{100} \times \sum_{i=1}^{n} (F_i \times W_i) \tag{10}$$

Table 2. Mapping between the Use Case Points and Business Continuity Points.

	Use Case Points	Business Continuity Points
Estimated complexity type	software complexity estimation	Business function recovery complexity estimation
Actors	Actors classified as Simple, average and complex, utilized to calculate the Unadjusted Actor Weight value (UAW)	Includes human and application level actors. each actor type is classified as simple, average and complex, utilized to calculate Unadjusted Human Weights (UHW) and Unadjusted Application Weights (UAPW) values
Use cases vs Business functions	Use cases are classified as simple, average and complex (according to the number of involved transactions), utilized to calculate unadjusted use case weights	Business functions are classified as simple, average and complex (according to the number of involved processes), utilized to estimate Unadjusted Business Function Weights (UBFW)
Unadjusted points estimation	Unadjusted Use Case Points: UCP = UAW + UUCW	Unadjusted Business Function Recovery Points: UBFRP = TUAW + UBFW
Technical factors	13 Technical factors (Limited Number)	Unlimited number of technical recovery factors
Environmental factors	8 Environmental factors (Limited Number)	Unlimited number of environmental recovery factors
Unexpected factors	No unexpected factors are considered	Unlimited number of unexpected recovery factors
Method of weight assignment	Based on the experience of IT project manager	Based on standard mathematical approach (Rank Order Centroid)
Adjusted points estimation	Adjusted Use Case Points (UPC)	Adjusted Business Function Recovery Points (ABFRP)

The above stated value will be considered towards the Recovery Time Effort (RTE) estimation in Man-Hours. A detailed mapping between the Use Case Points and the Business Continuity Points Method is depicted in Table 2.

The final step of the model includes the calculation of the Adjusted Business Function Recovery Points (ABFRP). The value will be provided by the multiplication of the Unadjusted Points value, the Technical Recovery Factors, the Environmental Recovery Factors and the Unexpected Recovery Factors according to the following formula:

$$ABFRP = UBFRP \times TRF \times ERF \times URF \qquad (11)$$

3.7   Business Function Recovery Scenarios

An important issue of the current work is the delineation of scenarios, according to which, the recovery procedure of a system or business function is influenced significantly, partly or slightly. As a consequence, 3 recovery scenario types should be considered for estimating the recovery complexity.

Simple Scenario. A *Simple Recovery scenario*, should be formulated under the following assumptions:

Human Level Actors: 1 Complex, 1 Average and 1 Simple
Application Level Actors in BF: 1 Complex, 1 Average and 1 Simple

$$\begin{aligned}
\text{Total Unadjusted Actor Weights} &= \text{UHW} + \text{UAPW} \\
&= 1*1.5 + 1*1 + 1*0.5 + 1*1.5 + 1*1 + 1 \\
&\quad *0.5 \\
&= 3 + 3 = 6
\end{aligned}$$

Business Processes in BF (consider number of activities): 1 Complex, 1 Average and 1 Simple

$$\text{Unadjusted Business Function Weights} = 1*1.5 + 1*1 + 1*0.5 = 3$$

And   Unadjusted Business Function Recovery Points $= 6 + 3 = 9$

Technical Recovery Factors: All technical factors which affect the process appear in their mildest form. Thus, the assessment value for each factor is 1. As a result, the Technical Recovery Factor should be equal to 0.75. Since *Environmental* and *Unexpected Factors* are calculated according to the same equation, their values should be also 0.75. The Total Score of Adjusted Business Function Recovery Points is:

$$\text{ABFRP} = 9*0.75*0.75*0.75 = 3.8$$

Average Scenario. An *Average Recovery scenario*, should be formulated under the following assumptions:

Human Level Actors in BF: 1 Complex, 2 Average and 2 Simple
Application Level Actors in BF: 1 Complex, 2 Average and 2 Simple

$$\begin{aligned}
\text{Total Unadjusted Actor Weights} &= \text{UHW} + \text{UAPW} \\
&= 1*1.5 + 2*1 + 2*0.5 + 1*1.5 + 2*1 + 2 \\
&\quad *0.5 \\
&= 4.5 + 4.5 = 9
\end{aligned}$$

Business Processes in BF: 2 Complex, 2 Average and 2 Simple

$$\text{Unadjusted Business Process Weights} = 2*1.5 + 2*1 + 2*0.5 = 6$$

And   Unadjusted Business Function Recovery Points $= 9 + 6 = 15$

Technical, Environmental and Recovery Factors: All factors have a medium impact on the recovery process for the given business function. Thus, the assessment value for each factor is 2.5. As a result, the Technical Recovery Factor should be equal to 1. Since *Environmental* and *Unexpected Factors* are calculated according to the same equation, their values should be also 1. The Total Score of Adjusted Business Function Recovery Points is:

$$ABFRP = 15 * 1 * 1 * 1 = 15$$

Complex Scenario. A *Complex Recovery scenario*, should be formulated under the following assumptions:

Human Level Actors in BF: 1 Complex, 3 Average and 3 Simple
Application Level Actors in BF: 1 Complex, 3 Average and 3 Simple

$$
\begin{aligned}
\text{Total Unadjusted Actor Weights} &= UHW + UAPW \\
&= 1 * 1.5 + 3 * 1 + 3 * 0.5 + 1 * 1.5 + 1 + 1 \\
&\quad * 0.5 \\
&= 6 + 6 = 12
\end{aligned}
$$

Business Processes in BF: 3 Complex, 3 Average and 3 Simple

$$\text{Unadjusted Business Process Weights} = 3 * 1.5 + 3 * 1 + 3 * 0.5 = 9$$

And $\text{Unadjusted Business Function Recovery Points} = 12 + 9 = 21$

Technical, Environmental and Recovery Factors: All factors which affect the process appear in their most severe form. Thus, the assessment value for each factor is 4. As a result, the Technical Recovery Factor should be equal to 1.25. In a similar way, the derived value for *Environmental* and *Unexpected Factors* is 1.25 as well. The Total Score of Adjusted Business Function Recovery Points is:

$$ABFRP = 21 * 1.25 * 1.25 * 1.25 = 41.01$$

3.8   Estimation of the Recovery Time (RT)

The Equation which should provide the Recovery Time (RT) value, is formulated after considering the derived results from the various recovery scenarios, as well as the RTO and MAO values assigned by the business continuity management. The RT value is provided by a quadratic function:

$$RT = 0.15 \times ABFRP^2 - 1 \tag{12}$$

The results of the calculations and the comparison between the estimated recovery time and the presently provided RTO and MAO values, are depicted in Table 3.

Table 3. Comparison between the estimated recovery time values by the Business Continuity Points and the currently proposed timeframes

Recovery scenario	URF	ABFRP	Recovery time (RT) (Hours)	RTO	MAO
Simple	0.75	3.8	~1.2	<2 h	2 h
Average	1	15	~32	<24 h	72 h
Complex	1.25	41.01	~251	<168 h	168 h

4 Discussion

In modern information age, the development of complex, distributed systems combined with organizational reliance upon on-line operations emphasized the importance of business continuity management which seeks to minimize the likelihood and magnitude of potential business interruptions, and encompasses Disaster Recovery Plans to guard against the major loss of IT services at any level in a system hierarchy [16]. Thus, the critical and challenging issue discussed in the current paper is whether the estimation of business function recovery complexity can be achieved by following a similar algorithm to the one utilized towards the estimation of software complexity. The author justifies this correlation following the statement of Laird and Brennan [17] according to which, complicated systems take longer to start and restart, which makes the outages longer.

So far, the available literature focuses mainly on the creation of business process complexity methods without any reference to time, or on methods that estimate execution time based solely on historical data. For instance, a study derived by Gruhn and Laue [18], analyze various software complexity metric models and map them to corresponding Business Process Management (BPM) metric models (Fig. 2). In this work the authors discuss how existing results in software complexity can be extended in order to analyze complexity of business process models. However, no information is provided towards the calculation of the time that is demanded in order to execute a process. Such models can be utilized as alternatives in order to determine the classification of the business function that is a sub-step of the overall recovery complexity estimation procedure. Another useful method presented by Ha, Reijers, J. Bae and H. Bae [19] analyzes the cycle time required for a process execution. However, the specific model includes 3 aspects of process information: process structure, resource capacity and statistical information. Contrary to this model, the method described in the current paper is designed in order to support prediction of the required time to execute the recovery process even if no past data is at the disposal of IT and Business Continuity Managers.

Another interesting element of the new method is that the TRF, URF and ERF values are independent from the number or factors and their ranking order listed in each category. This specific element is attributed to the efficient assignment of weight values according to the Rank Order Centroid Method, and permits the implementation of the model in a flexible manner. The validity of the model is checked, if after implementing simple, average and complex recovery scenarios, the estimated recovery time is reasonable, and within or close to the RTO and MAO values. It should be noticed that a

software complexity metric	corresponding metric for BPM	usage, significance
Lines of Code	number of activities	very simple, does not take into account the control-flow
cyclomatic number	CFC as defined by Cardoso	measures the number of possible control flow decisions, well-suited for measuring the number of test cases needed to test the model, does not take into account other structure-related information
max. / mean nesting depth	max. / mean nesting depth	provides information about structure, can be used complementary to the CFC metric
knot-count	number of handles	measure of "well-structuredness" (for example jumps out of or into control-flow structures) is always 0 for well-structured models can be used complementary to the CFC metric
cognitive weight	cognitive weight (tailored for BPM)	measures the cognitive effort to understand a model, can indicate that a model should be re-designed
(Anti)Patterns	(Anti)Patterns for BPM	experience with the patterns needed counting the usage of anti-patterns in a BPM can help to detect poor modeling
Fan-in / Fan-out	Fan-in / Fan-out	can indicate poor modularization

Fig. 2. Complexity metrics for software and business process models [18]

complex scenario exceeds the estimated MAO, which is equal to 168 h. The deviation is reasonable since the complex recovery scenario includes the presence of all the Unexpected Factors in their most severe form during the recovery process. In such circumstances, a reengineering of the recovery procedure may be considered by the IT and business continuity experts.

5 Conclusion – Future Work

The current paper analyzed a new theoretical approach for deriving efficient and timely IT business function recovery. The method is entitled Business Continuity Points, and is formulated by following the rules of software complexity estimation, as performed in the practically tested and scientifically acclaimed Use Case Points method. The method assumes that business function complexity estimation can be utilized as a driver to calculate the approximate time required to recover a business function. Existing models which estimate complexity for business process execution are utilized towards the understanding of the process, while another method that calculates time for the execution of the process, is based on statistical data. In contrast, the current method acts in a rather predictive manner and is aimed to estimate RTO and MAO values when no previous failure has occurred with regard to a specific business process. The estimated by the current model timeframes were compared to the business continuity timeframes proposed by the corresponding experts. The derived results ascertain the method's validity. Future work demands the practical implementation of the presented model. The author is currently working on its implementation to specific IT business functions in the Technical University of Liberec. A software tool based on VBA excel, is also under development in order to support automatic recovery time estimation.

Acknowledgments. The current work is supported by the SGS Project with the Number 21079, from Technical University of Liberec.

References

1. Martin, E.W., Brown, C.V., Hoffer, J.A., Perkins, W.C., Dehayes, D.W.: Managing Information Technology: What Managers Need to Know, 7th edn. Prentice Hall, Upper Saddle River (2011)
2. Rao, L., McNaughton, M., Osei-Bryson, K.-M., Haye, M.: The role of ontologies in disaster recovery planning. In: AMCIS 2009 Proceedings, Paper 713 (2009). http://aisel.aisnet.org/amcis2009/713
3. Miller, H.E., Engemann, K.J.: Using analytical methods in business continuity planning. In: Engemann, K.J., Gil-Lafuente, A.M., Merigó, J.M. (eds.) MS 2012. LNBIP, vol. 115, pp. 2–12. Springer, Heidelberg (2012)
4. Karner, G.: Resource estimation for objectory projects. In: Systems SF AB (1993)
5. Business Standard Institute: BS ISO 22301:2012 (2012)
6. Engemann, K.J., Henderson, D.M.: Business Continuity and Risk Management: Essentials of Organizational Resilience. Rothstein Associates, Brookfield (2012)
7. Information And Technology Services (ITS): Disaster Recovery/Business Continuity, University of Michigan (2014). http://www.mais.umich.edu
8. Gibson, D.: Managing Risks in Information Systems. Jones & Bartlett Learning, Burlington (2010)
9. National Institute of Standards and Technology-U.S. Department of Commerce: Contigency Planning Guide for Federal Information Systems, p. 16 (2010)
10. Snedaker, S.: Business Continuity and Disaster Recovery Planning for IT Professionals. Elsevier Inc., Burlington (2007)
11. Banerjee, G: Use Case Points -An Estimation Approach (2001)
12. Kusumoto, S., Matukawa, F., Inoue, K.: Estimating effort by use case points: method, tool and case study. In: 10th International Symposium on Software Metrics. IEEE Computer Society, Washington (2004)
13. Ochodek, M., Nawrocki, J., Kwarciak, K.: Simplifying effort estimation based on use case points. J. Inf. Softw. Technol. **53**, 200–213 (2011). Elsevier
14. Barron, F.H., Barrett, B.E.: Decision quality using ranked attribute weights. Int. Manag. Sci. **42**, 1515–1523 (1996)
15. Bagla, V., Gupta, A., Kukreja, D.: A qualitative assessment of educational software. Int. J. Comput. Appl. **36**, 1–7 (2011)
16. Caelli, W.J., Kwok, L.-F., Longley, D.: A business continuity management simulator. In: Rannenberg, K., Varadharajan, V., Weber, C. (eds.) SEC 2010. IFIP AICT, vol. 330, pp. 9–18. Springer, Heidelberg (2010)
17. Laird, M.L., Brennan, M.C.: Software measurement and estimation- a practical approach. IEEE Computer Society, Hoboken (2006)

18. Gruhn, V., Laue R.: Complexity metrics for business process models. In: 9th International Conference on Business Information Systems
19. Ha, B.-H., Reijers, H.A., Bae, J., Bae, H.: An approximate analysis of expected cycle time in business process execution. In: Eder, J., Dustdar, S. (eds.) BPM Workshops 2006. LNCS, vol. 4103, pp. 65–74. Springer, Heidelberg (2006)

Author Index